T0231257

Introduction to Light Emitting Diode Technology and Applications

Introduction to Light Emitting Diode Technology and Applications

GILBERT HELD

CRC Press
Taylor & Francis Group
Boca Raton   London   New York

CRC Press is an imprint of the
Taylor & Francis Group, an **informa** business
AN AUERBACH BOOK

Auerbach Publications
Taylor & Francis Group
6000 Broken Sound Parkway NW, Suite 300
Boca Raton, FL 33487-2742

© 2009 by Taylor & Francis Group, LLC
Auerbach is an imprint of Taylor & Francis Group, an Informa business

No claim to original U.S. Government works
Printed in the United States of America on acid-free paper
10 9 8 7 6 5 4 3 2 1

International Standard Book Number-13: 978-1-4200-7662-2 (Hardcover)

Library of Congress Cataloging-in-Publication Data

Held, Gilbert, 1943-
 Introduction to light emitting diode technology and applications / Gilbert Held.
      p. cm.
 Includes bibliographical references and index.
 ISBN 978-1-4200-7662-2 (alk. paper)
 1. Light emitting diodes. I. Title.

 TK7871.89.L53H45 2009
 621.3815'22--dc22   2008046202

Visit the Taylor & Francis Web site at
http://www.taylorandfrancis.com

and the Auerbach Web site at
http://www.auerbach-publications.com

Dedication

One of the advantages associated with living in a small town for almost 30 years is the commute to work. Having lived in New York City and the suburbs of Washington, D.C., moving to Macon, Georgia, provided me with over 10 hours per week of additional time that I could devote to writing manuscripts and preparing presentations. Over the past 30 years that I have lived in Macon, I was fortunate to be able to teach over 1,000 graduate students locally and perhaps 10,000 or more students who came to various seminars I taught throughout the United States, Europe, Israel, and South America. Many of those students were highly inquisitive and their questions resulted in a mental exercise for this veteran professor as well as second, third, and even fourth editions of some of the books I authored. In recognition of the students who made teaching truly enjoyable, this book is dedicated.

Contents

Preface

Light emitting diodes represent an old technology that has recently undergone numerous improvements that will result in its use being as ubiquitous as the cell phone. In fact, almost all cell phones today have their screens lit through the use of LEDs, which draw minimal power, a necessity when the primary purpose of the lightweight battery in a cell phone is to provide an extended operational time between recharges.

As you drive through a city, or examine the floor lighting on modern aircraft, or look for a flashlight, chances are excellent that you will encounter LED-based products. When you come to a traffic light and carefully look at the light you will note that the red, orange, and green lights are really made up of rows and columns of LEDs that form a matrix of a defined color. The use of LEDs results not only in a considerable savings in the use of electrical power, but, in addition, lowers maintenance costs. While LEDs do burn out, their life when used in a traffic light can extend considerably beyond 15 years and if one or two LEDs become inoperative there is no cause for alarm as the others keep on functioning as indicated in ads for the Energizer bunny.

In addition to traffic lights, LEDs are beginning to appear in high-end flashlights, as track lighting on airplane floors used to provide passengers with a guide to emergency exits, and even as replacements for florescent bulbs, which, in turn, had been developed as replacements for energy-inefficient incandescent light bulbs. Other applications for LEDs range from their use to transmit data over optical fiber to incorporation on different products to indicate the various operating modes of a device, such as "power on" indicated by a green or red LED, while other LEDs may be used to indicate the status of a different device function, such as a DVD recording a program to disk or copying a VHS tape to disk.

Today it is difficult, if not impossible, to get through our daily chores without coming into contact by either using or observing the use of LEDs. From their previously mentioned use in traffic lights and cell phones, to their use as power indicators on monitors and computers, they represent a truly ubiquitous technology. What is even more amazing is the fact that a considerable amount of development work continues to occur on LED technology that has resulted in several advances in the ability of the technology to support more efficient lighting and enhanced communications.

Because LEDs are closely associated with light, in addition to examining the evolution of the technology we will also focus our attention on the fundamentals of light, examining particle and wave theories, light metrics, visible and infrared light, how colored light occurs, and the effects of absorption, reflection, scattering and refraction of light. Doing so will provide a solid foundation for later chapters in this book that will cover LED basics, LEDs in lighting, LEDs in panels, LEDs used in optical communications, and other technologies.

As a professional author who has spent approximately 30 years working with different flavors of computer and optical technology, I welcome reader feedback. Please feel free to write to me in care of my publisher whose address is on the back cover of this book, or you might choose to send me an email to gil_held@yahoo.com. Because I periodically travel overseas, it may be a week or more until I can respond to specific items in the book.

Please feel free to also provide your comments concerning both material in this book as well as topics you may want to see in a new edition. While I try my best to literally "place my feet" in the shoes of the reader to determine what may be of interest, I am human and make mistakes. Thus, let me know if I omitted a topic you feel should be included in this book or if I placed too much emphasis on another topic. Your comments will be greatly appreciated.

Gilbert Held
Macon, Georgia

Acknowledgments

As the author of many books, a long time ago I realized that the publishing effort is dependent upon the work of a considerable number of persons. First, an author's idea concerning a topic must appeal to a publisher who is typically inundated with proposals. Once again, I am indebted to Rich O'Hanley at Taylor & Francis' CRC Press for backing my proposal to author a book focused upon a new type of Ethernet communications.

As an old-fashioned author who periodically travels, I like to use the original word processor—a pen and paper—when preparing a draft manuscript. Doing so ensures that I will not run out of battery power nor face the difficulty of attempting to plug a laptop computer into some really weird electric sockets I encounter while traveling the globe. Unfortunately, a publisher expects a typed manuscript and CRC Press is no exception. Thus, I would be remiss if I did not acknowledge the fine efforts of my wife, Beverly J. Held, in turning my longhand draft manuscript into the polished and professionally typed final manuscript that has resulted in the book you are now reading.

Once again, I would like to acknowledge the efforts of CRC Press employees in Boca Raton, Florida. From designing the cover through the editing and author queries, they double-checked this author's submission and ensured that it was ready for typesetting, printing, and binding. To all of you involved in this process, a sincere thanks.

1
INTRODUCTION TO LEDs

As you might expect, the purpose of an introductory chapter is to acquaint readers with the general topic of the book they are reading. Although this chapter is similar to such chapters in other books, due to the need to understand the fundamental aspects of light that are presented in Chapter 2 to appreciate light-emitting diode (LED) design, we will defer an in-depth description of LEDs until Chapter 3. In the interim, in this chapter, we will describe how a basic LED operates, obtain an overview of the basic technology associated with LEDs, examine how LEDs can be used in series and parallel circuits, note the use of resistors with LEDs, and understand how to develop circuitry that operates LEDs. In effect, we will return to an expanded prefix in this book by concluding this chapter with an overview of actual and potential LED applications and the advantages and disadvantages associated with their use. That said, perhaps you want to take a moment to grab your favorite drink and a few munchies as we turn our attention to the wonderful world of LEDs.

1.1 Basic Operation

The basic technology behind the development of the LED dates back to the 1960s when scientists were working with a chip of semiconductor material. That material was doped, or impregnated with impurities, to create a positive-negative or p-n junction.

1.1.1 The p-n Junction

Similar to a conventional diode, current will flow from the p-side of a semiconductor to its n-side, but not in the reverse direction. The

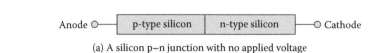

(a) A silicon p–n junction with no applied voltage

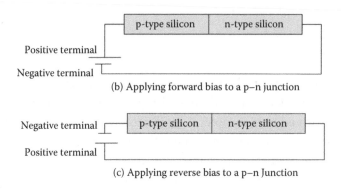

(b) Applying forward bias to a p–n junction

(c) Applying reverse bias to a p–n Junction

Figure 1.1   The p-n semiconductor junction.

p-side is also referred to as the *anode*, and the n-side is also known as the *cathode*.

Figure 1.1 contains a series of illustrations that indicate how a basic diode is formed and represents the forerunner or predecessor of the LED. Thus, one common term for LED is "a son of a diode."

1.1.1.1 No Applied Voltage   At the top of Figure 1.1a, a silicon p-n junction with no applied voltage is shown. Both p- and n-doped semiconductors are relatively conductive; however, the junction between them is a nonconducting layer that is commonly referred to as the *depletion zone*. The depletion or nonconducting area or zone occurs when the electrically charged carriers in the doped n-type silicon (referred to as *electrons*) and p-type silicon (referred to as *holes*) attract and eliminate one another in a process referred to as *recombination*. Through the manipulation of the nonconductive layer between the p- and n-type silicon, a diode can be formed. The resulting diode forms an electrical switch that allows the flow of electricity in one direction but not in the opposite direction.

1.1.1.2 Applying Forward-Bias   In Figure 1.1b, a positive terminal is shown connected to the anode, and the negative terminal is connected to the cathode. The result of this connection is a forward bias, which pushes the holes in the p-region and the electrons in the n-region toward the junction, in effect reducing the width of the depletion zone. That is, the positive charge applied to the p-type silicon repels the holes from the n-type silicon, whereas the negative charge applied to the n-type silicon repels the electrons from the p-type silicon. The net effect of the positive and negative terminal connections is to push the electrons and holes toward the p-n junction, lowering the barrier potential required to reduce the nonconducting depletion zone so that it becomes so thin that charge carriers in the form of electrons can tunnel across the barrier p-n junction by increasing the forward bias voltage. Thus, electrons begin to enter the p-type silicon and move from hole to hole through the crystal, making it possible for electric current to flow from the negative terminal to the positive terminal of the battery.

1.1.1.3 Applying Reverse-Bias In Figure 1.1c, the polarity of the battery connection is reversed, resulting in a reverse-bias effect. That is, the p-type region is now connected to the negative terminal of a power supply, which results in the holes in the p-type silicon being pulled away from the p-n junction. In effect, this action results in increasing the width of the nonconducting depletion zone. Because the n-type silicon is connected to the positive terminal, this action also results in the electrons being pulled away from the junction, which widens the barrier and significantly increases the potential barrier, which in turn increases the resistance to the flow of electricity. Thus, a reverse-bias connection minimizes the potential for electric current to flow across the p-n junction. However, as the reverse voltage increases to a certain level, the p-n junction will break down, allowing current to begin to flow in the reverse direction. This action is associated with the use of Zener or avalanche diodes. From the preceding text, it is clear that a p-n junction of silicon can be used as a diode, enabling electric charges to flow in one direction through the junction but not in the opposite direction unless a very high voltage potential is used in a reverse-bias condition. When used in a positive bias, negative charges in the form of electrons easily flow from n-type material to p-type material, whereas the reverse is true for holes. However, when the p-n junction is reverse biased, the junction barrier is widened, which increases the resistance to the flow of current. Now that we have a general appreciation for how a diode operates, let's turn our attention to the basic operation of an LED.

1.1.2 *LED Operation*

In Section 1.1.1 of this chapter, we examined the operation of the p-n junction, which is common to diodes and LEDs. In the following sections, we will examine how an LED generates light via the use of doping material, before turning our attention to a short description of the evolution of the LED.

1.1.2.1 Similarity to a Diode   An LED can be considered to resemble a diode because it represents a chip of semiconducting material that is doped or impregnated with impurities to form a p-n junction. Similar to a diode, current easily flows from the p-side to the n-side of the semiconductor via a forward-bias potential, but not in the reverse direction.

1.1.2.2 Crossing the Barrier   When an electron crosses the barrier and meets a hole, it falls into a lower energy level and releases energy in the form of a *photon*. The photon is a carrier of electromagnetic radiation of all wavelengths. The actual wavelength of light generated and its color that corresponds to the emitted wavelength is dependent on the band gap energy of the materials used to form the p-n junction. For example, for silicon or germanium diodes, the electrons and holes combine via a forward-bias voltage such that a nonradiative transition occurs, which results in no optical emission as the semiconductors represent indirect band-gap material. However, through the initial use of gallium arsenide and other materials, a direct band gap with energies corresponding to near-infrared, visible, or near-ultraviolet light could be generated by the evolving LED.

1.1.3 *LED Evolution*

In the following sections we will briefly discuss the evolution of the LED. This discussion will include how experiments in the use of different doping materials resulted in the development of different colors and color intensities for LEDs.

1.1.3.1 The First LED The actual invention of the first practical LED is attributed to Nick Holonyak in 1962. Holonyak, who attained the position as the John Bardeen Professor of Electrical and Computer Engineering and Physics at the University of Illinois, was the first student of Professor John Bardeen, who was one of the inventors of the basic transistor during the 1950s. After completing graduate school in 1954, Nick Holonyak took a job with Bell Laboratories and contributed to the development of the integrated circuit. Later, while working at General Electric, Holonyak was responsible for the development of the p-n–p-n switch, which is now widely used in homes and apartments as a dimmer switch to control lighting to a chandelier on another light source.

On April 23, 2004, Mr. Holonyak was officially recognized as the inventor of the LED at a ceremony that was held in Washington, D.C. At that ceremony, Holonyak received the half-million dollar Lemelson-MIT Prize for Invention, which is the world's largest cash prize awarded to an inventor.

1.1.3.2 Doping Materials Although Nick Holonyak is recognized as the inventor of the LED, during the 20th century, several companies either inadvertently or by design were able to generate electroluminescence from different materials by the application of electric fields. For example, in a report (1923), the generation of blue electroluminescence was based on the use of silicon carbide (SiC) that had been manufactured as sandpaper grit. Although the sandpaper grit inadvertently contained what are now referred to as p-n junctions, at the time the generation of light was both poorly controlled and not exactly scientifically understood. However, fast-forwarding to the 1960s, SiC films were prepared by a much more careful process than manufacturing sandpaper grit, whereas the evolution of p-n junction semiconductors was driven by curiosity and practical experimentation. In fact, by the mid-1960s this author remembers taking several graduate physics courses that involved the doping of various materials to create p-n semiconductor junction diodes. By the later portion of the 1960s, p-n junction devices were fabricated that resulted in the development of blue LEDs. Although this first generation of blue LEDs were extremely inefficient, subsequent efforts to improve the efficiency of blue SiC LEDs only marginally improved due to an indirect band gap in the p-n junction. By the early 1990s, the maximum efficiency of blue SiC LEDs that emitted blue light at a 470 nm wavelength was only approximately 0.03 percent. Thus, the low efficiency of SiC LEDs resulted in scientists turning their attention to other semiconductor materials both as a mechanism to enhance efficiency as well as a method to generate light from other areas of the frequency spectrum. One such approach was the development of infrared LEDs based on the use of GaAs.

1.1.3.2.1 Gallium Arsenide LEDs   During the 1960s, infrared (IR) LEDs were developed based on the use of GaAs that was grown as a crystal, then sliced and polished to form the substrate of a p-n junction diode. As previously mentioned, the use of GaAs resulted in the development of IR LEDs whose application capability was limited owing to the absence of visible light.

The development of IR LEDs resulted in several key differences between the electrical characteristics of IR and visible LEDs. Those differences are primarily in the forward voltage used to drive the LED, its rated current, and the manner in which its output is rated. IR LEDs typically have a lower forward voltage and higher rated current than a visible LED due to the material properties of the p-n junction. Concerning their output rating, because IR LEDs do not output light in the visible spectrum, they are commonly rated in milliwatts. In comparison, the output of visible LEDs is rated in millicandelas (mcd), where 1000 mcd equals a candela, which represents lumens divided by the beam coverage. In Chapter 2 when we discuss the fundamentals of light, we will also describe various light-related terms as well as techniques associated with measuring the light output.

1.1.3.2.2 Gallium Arsenide Phosphide LEDs   To obtain a visible light emission, GaAs was alloyed with phosphide (P), resulting in a gallium arsenide phosphide (GaAsP)-based LED that emitted red light.

1.1.3.2.3 Use of Other Doping Materials   During the 1960s, scientists and physicists experimented with the use of various doping materials to generate various portions of the visible wavelength. The doping of GaP with nitrogen resulted in the generation of a bright yellow green

0.550 nm wavelength, whereas at RCA's then central research laboratory in Princeton, New Jersey, the use of gallium nitride (GaN) was used to generate blue light peaking at a wavelength of 475 nm during the summer of 1971. Approximately a year later, Herbert Maruska at RCA decided to use magnesium as a p-type dopant instead of zinc. Maruska then began growing magnesium-doped GaN films, resulting in the development of a bright violet-colored LED emitting light at 430 nm.

Due to RCA's financial problems during the mid-1970s, work on a blue LED using GaN was cancelled. However, in 1989, Isamu Akasaki was able to use magnesium-doped GaN to achieve conducting material by using an electron beam annealed magnesium-doped GaN. A little more than a decade later, in 1995, a blue and green GaN LED with an efficiency exceeding 10 percent was developed at Nichia Chemical Industries in Japan.

1.1.3.2.4 Rainbow of Colors Over a period of approximately 50 years, LEDs have been manufactured using different inorganic semiconductor materials to generate a wide variety of colors. Table 1.1 lists in alphabetical order common semiconductor materials used to create LEDs as well as the type of generated light. Note that the use of certain types of semiconductor materials is currently under

Table 1.1 Use of Semiconductor Materials to Generate LED Light

SEMICONDUCTOR MATERIALS	LED EMISSION
Aluminum gallium arsenide (AlGaAs)	Red and infrared
Aluminum gallium phosphide (AlGaP)	Green
Aluminum gallium indium phosphide (AlGaInP)	Bright orange red, orange, yellow
Aluminum gallium nitrate (AlGaN)	Near to far ultraviolet
Aluminum nitrate (AlN)	Near to far ultraviolet
Diamond (C)	Ultraviolet
Gallium arsenide phosphide (GaAsP)	Red, orange and red, orange, yellow
Gallium phosphide (GaP)	Red, yellow, green
Gallium nitrate (GaN)	Green, emerald green
Gallium nitrate (GaN) with AlGan quantum barrier	Blue, white
Indium gallium nitrate (InGaN)	Bluish green, blue, near ultraviolet
Sapphire (Al$_2$O$_3$) as substrate	Blue
Silicon (Si) as substrate	Blue (under development)
Silicon carbide (SiC)	Blue
Zinc selenide (ZnSe)	Blue

development. This development effort is primarily focused on research into generating bright white light. Due to the development of several methods to generate bright white light, the number of applications available for LEDs has considerably expanded, including one application familiar to many consumers. That application is the use of bright white LEDs in high-end flashlights.

1.1.4 Voltage and Current Requirements

As indicated earlier in this chapter (Section 1.1.2.1), an LED has the electrical characteristics of a diode. This means that it will pass current in one direction but block it in the reverse direction. Depending on the semiconductor material and its doping, the LED will emit light at a particular wavelength.

In general, LEDs require a forward operating voltage of approximately 1.5–3 V and a forward current ranging from 10 to 30 mA, with 20 mA being the most common current they are designed to support. Both the forward operating voltage and forward current vary depending on the semiconductor material used. For example, the use of gallium arsenide (GaAs) with a forward voltage drop of approximately 1.4 V generates infrared to red light. In comparison, the use of gallium arsenide phosphide (GaAsP) with a voltage drop near 2 V is used to generate wavelengths that correspond to frequencies between red and yellow light, whereas gallium phosphide LEDs have a blue-green to blue color and a voltage drop of approximately 3 V.

1.1.4.1 Manufacture of LEDs
In a manufacturing environment, different amounts of arsenide and phosphide are commonly used to produce LEDs that emit different colors. Currently, blue and bright white LEDs are more difficult to manufacture and are usually less efficient than other LEDs. Their lower efficiency and greater manufacturing difficulty results in an increase in their unit cost.

LEDs are manufactured in several sizes and shapes. Some are manufactured as multicolor devices that contain both a red and a green chip, enabling the production of light between the two colors. Tricolor, red, blue, and green (RGB), LEDs are also manufactured as well as various types of white LEDs that vary in intensity and are used for different applications. Applications of LEDs range from use as indicators to lighting and data transmission. Visible light LEDs are primarily used for indicator lights, such as an emergency path on an aircraft floor. In comparison, high-intensity white LEDs are used for short-range lighting in flashlights, whereas IR LEDs are commonly used for data transmission. Later in this chapter, we will describe and discuss a range of LED applications that make the device as ubiquitous as the pen.

1.1.4.1.1 LED Legs The general fabrication process that results in the manufacture of LEDs is so well thought out that it becomes difficult to use them incorrectly. LEDs are manufactured with two "legs" protruding from the flat edge of the device, as illustrated in Figure 1.2. On modern LEDs, the anode (+) is longer than the cathode (–), with the latter marked by a flat edge. Although the anode is marked with the letter "a," a "c" or "k" is used to mark the cathode, with the letter "k" more frequently used. Unfortunately, older LEDs were not explicitly fabricated, and often their improper connection resulted in the device burning out.

Returning our attention to Figure 1.2 note that the emitted light is reflected off the plastic case at different angles and, unlike a laser, is not coherent light.

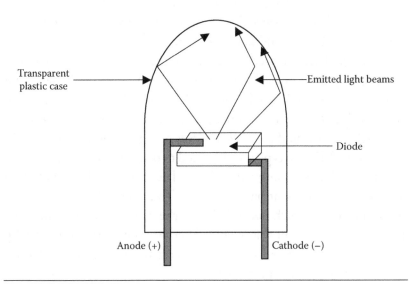

Figure 1.2 LED fabrication.

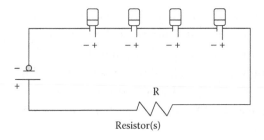

Figure 1.3   Placing LEDs in series.

1.1.4.2 Parallel and Series Operations   Similar to other electronic devices, LEDs can be used in two basic types of circuits: series and parallel.

1.1.4.2.1 Series Operations   A number of LEDs placed in series is similar to Christmas lighting. That is, if one should fail, it will result in an open circuit that stops the flow of current to other devices beyond the failed device. An exception to this are LEDs whose failure enables current to bypass the failed device, allowing the other LEDs to continue to illuminate.

Figure 1.3 illustrates the connection of four LEDs in series with one another driven by a 12 V power source. Note that the LEDs are positioned such that the cathodes (–) and anodes (+) alternate in their connection to the wiring that forms the circuit. Otherwise, placing two anodes (+) or two cathodes (–) in sequence would disable the circuit and the LEDs would not illuminate.

With four LEDs placed in series using a 12 V power source, the voltage going through each LED is 12/4 or 3 V. If you only had three LEDs in series, each would receive 12/3 or 4 V. Similarly, if there were two LEDs in series and the power source continued to be 12 V, then each LED would have 12/2 or 6 V going through the device.

Because LEDs are typically designed to operate between 2 and 4 V, too much voltage passing through the LED can result in its failure as well as an unpleasant burning smell. To prevent this, it's common to add a resistor, which not only limits the voltage drop but, in addition, limits the current that would otherwise flow through the LED. Shortly, we will discuss the use of resistors in more detail, including computation of their value in ohms.

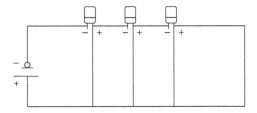

Figure 1.4 LEDs operating in parallel.

1.1.4.2.2 Parallel Operations   In addition to connecting LEDs in series, they can also be connected in parallel. Figure 1.4 illustrates the connection of three LEDs in parallel to a common power source. Note that each parallel circuit has at least two independent paths in the circuit that provide a return to the source.

When two or more LEDs are connected in parallel, they have the same potential difference (voltage) across their ends. Thus, a 3 V power source would result in the same amount of voltage received by each LED.

There are a few general restrictions associated with using LEDs in parallel. First, the LEDs need to have the same voltage rating, which usually implies that they have the same color. This is because electricity flows along the path of least resistance, which means that the LEDs that require less power would illuminate whereas the ones requiring more power would remain unlit. To fix this problem, you would need to insert a resistor into each parallel circuit to, in effect, equalize all the LEDs. Figure 1.5 illustrates the use of three resistors whose values for now we will refer to as x, y, and z Ω.

As an alternative to the use of three resistors, it's also possible to use a single resistor. Thus, Figure 1.5 could be redrawn as Figure 1.6. However, when using a common resistor, you should use LEDs with

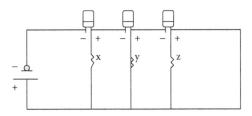

Figure 1.5 Placing LEDs in parallel.

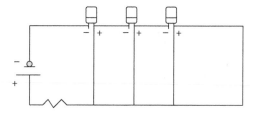

Figure 1.6   Placing LEDs in parallel using a single resistor.

the same voltage and amperage ratings, a point that we will discuss later in this chapter.

1.1.4.3 Current Limitation Considerations   LEDs are designed to operate at a relatively low level of current, such as 20 mA. Thus, applying a voltage directly to a single LED or grouping of LEDs can result in LED burnout and even a potential explosion.

To limit the current flowing through one or more LEDs, the use of a resistor is required. Because LEDs can be connected in either series or parallel, let's review the general operation of each type of circuit.

1.1.4.3.1 Series Circuit   An LED series circuit is a circuit in which LEDs are arranged one after the other in a chain. Thus, the current has only one path to take through the circuit. When a number of LEDs are connected in series, only one resistor is required to be inserted into the circuit. However, if you do not have a resistor with the correct value, you can use multiple resistors because the total resistance in a series circuit is the sum of the resistance of each resistor. Figure 1.7 illustrates a series circuit that contains three resistors, labeled R_1, R_2, and R_3. R_1 is 10 Ω, R_2 is 20 Ω, and R_3 is 30 Ω; thus, the total resistance is 60 Ω. This means that if you need 60 Ω of resistance, you could use one 60 Ω resistor, two 30 Ω resistors, or another combination that adds up to 60 Ω.

Using Ohm's law, where $V = IR$, the total current that will flow in the circuit becomes

$$I = V/R$$

Thus, as the resistance increases and the voltage remains constant, the current decreases. To illustrate this, let's assume the voltage is provided by a 12 V battery and the resistance is 60 Ω. Then, the current that

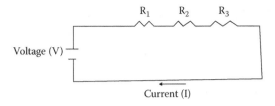

Figure 1.7   A series circuit with three resistors.

flows in the circuit is 12/60 or 0.2 A or 200 mA. Because most LEDs support a much lower current, you would need to use a larger resistor.

1.1.4.3.1.1 Computing the Resistor Value   The key to the correct operation of an LED is obtaining the correct resistor. Otherwise, a small change in the voltage across an LED can result in a large change in current that can literally fry the LED. This is because the relationship between LED voltage and current is defined by the device's operating curve, and an LED's rating, for example, 3.2 V @ 20 mA, represents one point along its operating curve. Thus, it's more useful to consider driving an LED with a current of a given value instead of applying a voltage. This is because knowing the voltage across an LED does not allow you to determine the current flowing through the device unless it's being operated at a particular point along its operating curve. Thus, through the use of a resistor, which provides a linear relationship between voltage and current, you can easily control the current flowing through an LED.

The formula used to determine the resistor value required for one or more LEDs placed in a series circuit is as follows:

$$R = (V_s - N \times V_f)/I$$

where

R = resistance in ohms

V_s = supply voltage in volts

N = number of LEDs placed in series

V_f = forward voltage required by each LED in volts (commonly listed on the LED package)

I = maximum current rating in amps (commonly listed on the LED package)

Note that this equation represents Ohm's law as the voltage across the resistor is $V_s - NV_f$. For most LEDs, V_f is commonly around 2 V; however, for blue and white LEDs, the required voltage is approximately doubled.

To illustrate the use of the preceding equation, let's assume you want to illuminate four white LEDs, each with a forward voltage of 2.5 V and a maximum current rating of 20 mA. Using the preceding equation, we obtain

$$R = (12 \text{ V} - 4 \times 2.5 \text{ V})/0.020 \text{ A} = 100 \text{ }\Omega$$

Thus, you would need a resistor of at least 100 Ω to maintain a level of current through the LEDs to prevent their burnout.

To assist users who may be thinking about implementing one or more LEDs in different products, several Web sites include resistor calculators. Figure 1.8 illustrates the use of a JavaScript program (http://www.quickar.com/noqbestledcalc.htm) to compute resistance. Note that the results are the same as our prior computation; however, the manner in which the computation occurs is not addressed by the indicated Web site.

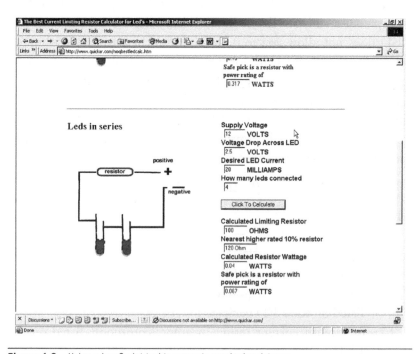

Figure 1.8   Using a JavaScript tool to compute required resistance.

1.1.4.3.2 Parallel Circuit A parallel circuit, as its name implies, is a circuit in which LEDs, resistors, and other devices are arranged parallel to one another. Although the voltage across each parallel circuit is the same, the current breaks up, with some flowing along each parallel branch and then recombining when the branches meet.

Figure 1.9 illustrates a three-branch parallel circuit. In this example, I_1, I_2, and I_3 represent currents flowing through the branches, which have resistances R_1, R_2, and R_3.

The total resistance for a series of resistors in parallel is computed by adding up the reciprocals of each resistor and then taking the reciprocal of the total. That is, the total resistance (R_T) becomes

$$\frac{1}{R_T} = \frac{1}{R_1} + \frac{1}{R_2} + \frac{1}{R_3}$$

or

$$R_T = 1/(1/R_1 + 1/R_2 + 1/R_3)$$

To illustrate the computation of total resistance, assume R_1 and R_2 are each 20 Ω, and R_3 is 10 Ω. Then,

$$R_T = 1/(1/20 + 1/20 + 1/10) = 5 \ \Omega$$

If the battery shown in Figure 1.9 is 12 V, then by Ohm's law, the total current flowing in the circuit becomes $I = V/R$ or 12/5 = 2.4 A. The individual currents flowing in each branch can be computed similarly. That is, $I_1 = 12/20$ or 0.6 A, $I_2 = 12/20$ or 0.6 A, and $I_3 = 12/10$ or 1.2 A. Note that collectively, $I = I_1 + I_2 + I_3$, which is the total current flowing through the circuit.

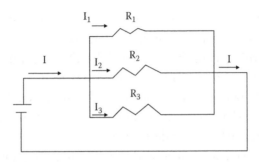

Figure 1.9 Parallel circuit operation.

1.1.4.3.2.1 *Determining the Resistor Values*   If you need to hook up LEDs in parallel, you would analyze the branches of the circuit based on the fact that the supply voltage remains the same for each branch. Then, you would use the forward voltage of each LED and its maximum current rating to determine the value of each resistor to be used in each branch of the circuit. For example, let's assume you have a 6 V battery that will drive three LEDs in parallel. Let's further assume that each of the three LEDs has a forward voltage of 2.3 V. Finally, let's assume each LED has a 20 mA current rating.

For each branch, the 6 V power source is reduced by 2.3 V associated with the LED, resulting in a voltage potential of 3.7 V on the branch. Because the LED is rated at 20 mA, then from Ohm's law, we need a resistor for each of the three branches, which we compute as follows:

$$R = \frac{3.7 \text{ V}}{20 \text{ mA}} = 185 \text{ } \Omega$$

Because $1/R_T = 1/R_1 + 1/R_2 + 1/R_3$, the total resistance in the three-branch circuit becomes

$$1/R_T = 1/185 + 1/185 + 1/185 = 61.66 \text{ } \Omega$$

Figure 1.10 illustrates the LED calculator program located at http://www.hebeiltd.com.cn/?p=zz.led.resistor.calculator that can be used to simplify the determination of resistance when your LEDs are to operate in parallel. In fact, if you enter the search term "LED calculator," you can find dozens of Web sites that provide an assortment of LED calculator tools that can simplify your computations or serve as a double check when you do the computations manually.

1.1.4.3.2.2 *Using a Shared Resistor*   Although it's acceptable to connect LEDs in series or parallel, a few words of caution are warranted concerning connecting several LEDs in parallel using one shared resistor as shown in Figure 1.11. This (the use of a shared resistor) may work, but often the results may be unexpected. In addition, it's also possible that the configuration shown in Figure 1.11 can result in one or more LEDs being destroyed.

The key problem associated with using a single resistor with LEDs in parallel will occur when the LEDs have different voltage ratings.

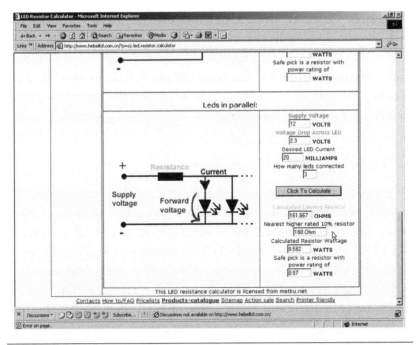

Figure 1.10   Using a Web-based tool to compute resistance when LEDs are connected in parallel.

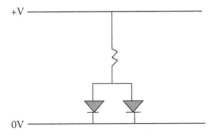

Figure 1.11   Avoiding the use of a common resistor.

When this occurs, only the LED with the lowest voltage will illuminate, and it's possible that it could be destroyed as a result of a large
current flowing through it.

  If you need to use a common resistor for two or more LEDs in
parallel, then you should use identical LEDs. This will ensure that
each LED illuminates, and eliminate the potential for LED burnout.
Although resistors are relatively inexpensive, sometimes space constraints can necessitate using a single resistor with two or more LEDs

in parallel. However, when LEDs are used in parallel, if possible, each one should have its own resistor.

If you carefully look at Figure 1.11, you will note that this author used a common symbol to illustrate the two LEDs in parallel. That symbol is a triangle with a line across its output, the latter being used sometimes to indicate light emission.

1.2 Types, Functions, and Applications

To conclude this chapter, we will turn our attention to several LED topics that will expand our knowledge about this ubiquitous electronic device. First, we will focus our attention on the different types of LEDs based on their physical characteristics and color generation capability. Once this is accomplished, we will examine the functions and applications associated with the use of LEDs. However, because LEDs are ubiquitous, our examination of applications will be limited to describing the major areas that use this device. In actuality, the use of an LED is only limited by one's imagination.

1.2.1 Types of LEDs

In the following sections, we will turn our attention to the general physical characteristics and color generation capability of LEDs.

1.2.1.1 Physical Characteristics When work commenced on developing semiconductor materials and doping the materials to generate light, the resulting LED was far from being miniaturized. Gradually, efficiencies in the development of manufacturing small disk drives, modems, personal digital assistants (PDAs), and other electronic devices with relatively small footprints facilitated a revolution in the development of LEDs. Today, LEDs can be obtained in a wide variety of shapes and sizes. LEDs can be obtained in round, square, rectangular, and triangular cross-sectional shapes. The most common shape is a round cross-section LED, as it's easy to install by simply drilling a hole matching the size of the LED diameter into the surface of the device it is to be mounted on. Then, a spot of glue can be used to fasten the LED to the surface of the device. In addition to glue, some LEDs are designed to be pressed into a clip to facilitate their attachment to a device. Initially, LEDs were individually encased in a plastic housing that had leads for the anode and cathode protruding from the bottom. This design, although still in use, represents only a small portion of currently manufactured LEDs due to the development of surface mount LEDs. However, before discussing surface mount LEDs, a few words are in order concerning the life expectancy of an LED.

1.2.1.1.1 Life Expectancy   One of the major advantages of an LED is its life expectancy. Most modern LEDs have a half-life of approximately 100,000 hr prior to its brightness level being halved. Because a year consists of 8760 hr, this means that the half-life of an LED is approximately 11.4 years, which explains why they have become the preferred lighting source for traffic lights and other applications where it is difficult to predict bulb burnouts and even more troublesome if a bulb used in an application fails.

1.2.1.1.2 Surface Mount LEDs   Today, perhaps the most popular type of LED is the surface mount device (SMD). An SMD LED represents an integrated LED as an epoxy package, which facilitates its use as an indicator in denoting the operational status of a device. The epoxy packaging provides a focus for the LED light beam, otherwise the resulting beam would have a wider viewing angle but would not be as visible.

1.2.1.1.2.1 Sizes   SMD LEDs are available in four popular sizes. These sizes are designated by the use of four numeric codes. Table 1.2 lists the SMD LED designators and their package sizes in terms of length, width, and height in millimeters. As indicated in the table, the 0402 SMD LED represents the smallest package whereas the 1206 represents the largest.

Table 1.2   Surface Mount Device (SMD) LEDs

DESIGNATOR	LENGTH × WIDTH × HEIGHT
0402	1.0 mm × 0.5 mm × 0.45 mm
0603	1.6mm × 0.8 mm × 0.60 mm
0805	2.0 mm × 1.25 mm × 0.80 mm
1206	3.2 mm × 1.5 mm × 1.10 mm

Because there are 2.54 cm in an inch, a centimeter is approximately 0.39 in. in length. As a millimeter is a tenth of a centimeter, then the length of a millimeter is equal to approximately 0.039 in. Thus, if you use a ruler and measure the letters in the inscription "United States of America One Dime" on a 10¢ U.S. coin, you would note that the 0603 SMD LED is approximately the size of the letter "D" in "Dime," whereas a 0805 SMD LED would cover the letter "I."

SMD LEDs are similar to standard LEDs, having a typical forward current of 35 mA with an average forward voltage of 3.6 V and a maximum forward voltage of 4.0 V. Because SMD LEDs are relatively tiny, they are normally packaged within a tape reel to facilitate their use. Once removed from tape, the SMD LED can be easily soldered to a circuit board or into the housing prefabricated on a disk drive, monitor, modem, or another device, where its illumination indicates a predefined action or activity. For example, a green LED might be used to indicate that a monitor was in a powered-on state.

1.2.1.2 Colors The actual color generated by an LED is determined by the semiconductor material and its doping, and not by the color of the plastic body that forms an LED package. Today, LEDs are available in a variety of colors, ranging from red, orange, and yellow (the "ROY" in the famous name ROY G. BIV used as a mnemonic to remember primary colors), to amber, green, blue, and white. LEDs can be obtained in uncolored packages that can be diffused (milky) or clear. Colored packages are also offered as diffused or transparent.

1.2.1.2.1 Color Variations Through the use of multiple LEDs, it becomes possible to obtain a bicolor or tricolor LED. A bicolor LED is formed by packaging two LEDs that are wired in an inverse parallel combination; that is, one is wired backward, enabling one of the LEDs to be illuminated at one time, depending on the lead on which voltage is applied.

When two LEDs are combined in one package with three leads, the result is a tricolor LED. The name "tricolor" results from the fact that the light generated by each LED can be mixed to form a third color when both LEDs are turned on.

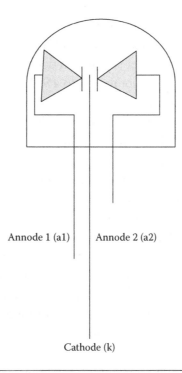

Annode 1 (a1)   Annode 2 (a2)

Cathode (k)

Figure 1.12   A tricolor LED.

Figure 1.12 illustrates a tricolor LED. Note that the center lead represents a common cathode (c) for both LEDs, whereas the outer leads represent the anodes (a_1, a_2) for each LED. Thus, each LED can be illuminated separately, or both can be lit to form a third color, for example, mixing red and green to obtain yellow.

1.2.1.3 Flashing LEDs   You may have noticed that when certain types of electronic devices are turned on, the LED flashes on and off repeatedly until you turn power off. One device this author has used for years that has a flashing LED is an ionizer dog brush. As our family Shitzu looks at the flashing light, perhaps the manufacturer of the device incorporated it to gain the attention of the animal being brushed. Unfortunately, whether the LED is flashing or turned off, it's still rather difficult to brush the family dog.

A flashing LED consists of an integrated circuit (IC) and an LED. The purpose of the IC is to flash the LED by turning power to the diode on and off at a fixed rate, typically 3 or 4 flashes per second (3 Hz to 4 Hz). A flashing LED package, including the IC, is designed to be directly connected to a power supply and does not require the use of a resistor. Typically, they are connected to a 9–12 V power source.

By combining a bicolor or tricolor LED with an IC, some manufacturers offer combined devices. For example, you can obtain an RGB flashing LED that toggles from red to blue to red to green to red, repeating this pattern over and over when power is applied to the device.

Through the use of one or more IC, several LED effects can occur when bicolor or tricolor LEDs are integrated with the ICs. For example, by removing power from one anode while applying power to the other anode of a tricolor LED, the color generated can appear to fade to a second color. Due to the visual attraction resulting from flashing LEDs, they have found a viable market, being incorporated into a range of products from toys to women's bras. Later in this chapter, we will discuss in more detail the range of LED applications and some of the products that use this ubiquitous device.

1.2.1.4 LED Displays In concluding our brief overview of the various types of LEDs, this author would be remiss if he did not discuss their packaging or grouping to form a variety of displays, ranging from counters and timers to a matrix that can be illuminated to show numbers, letters, and various types of graphics. The creation of many LED displays are obtained through the use of dot matrix, 7-segment, starburst, and similar LED component packages. Figure 1.13 illustrates the 7-segment, starburst, and dot matrix component display packages.

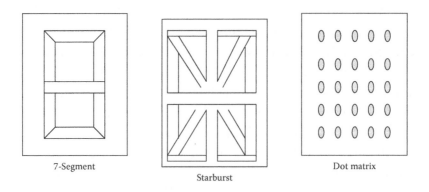

7-Segment   Starburst   Dot matrix

Figure 1.13 Common LED component packages.

In examining the LED component packages illustrated in Figure 1.13, note that the 7-segment package can be used to illuminate a number between 0 and 9. Thus, four 7-segment components are commonly used in clock radios, whereas additional 7-segment components are grouped together in a common housing with applicable circuitry to create digital counters and timers.

If you carefully examine a modern digital clock radio, you will probably note the use of several types of LEDs. In addition to four 7-segment components used to display the time, there are separate LEDs to indicate an alarm setting as well as a backlight that may be connected to a switch, which enables you to dim the display.

The starburst LED component provides the ability to illuminate from 1 to 15 lines. This component can be considered to represent an enhancement beyond the 7-segment display capabilities, as it can be programmed to illuminate numerics as well as certain graphics.

The third component shown in Figure 1.13, the dot matrix module, can be obtained in several configurations, such as 5 rows × 5 columns and 7 rows × 5 columns. By illuminating one or more dots, numbers, graphics, and symbols can be displayed.

1.2.2 Applications

The primary function of an LED is to provide a source of illumination at a defined wavelength. By grouping LEDs together in some fashion, it becomes possible to create counters, clocks, and other types of digital displays. Through the use of ICs, it becomes possible to flash LEDs in a predefined sequence, resulting in a visual attention-generating addition to many products, ranging from toys, cups, and safety band to women's bras.

1.2.2.1 Lighting With the development of white-light-generating LEDs, they have found a relatively new series of lighting applications. For example, if you examine modern high-end flashlights, you will note that the vast majority now use LEDs. Not only do they require considerably less power but, in addition, the life of an LED is typically several orders of magnitude beyond that of the bulb used in conventional flashlights. Other lighting applications in which LEDs are used include traffic lights and architectural lighting; they are even beginning to replace conventional incandescent lightbulbs. Concerning the latter, currently LED lightbulbs are relatively expensive, with several publications containing advertisements during 2008 that indicated a cost per bulb between $25 and $100. Although such bulbs are currently a niche market for use in hard-to-reach areas that make the long life of the bulb cost-effective, within the next decade we can expect prices of LED bulbs to significantly decrease. When this occurs, their extremely low power requirements, long life, and the fact that they produce nearly no heat will make them attractive replacements for incandescent lightbulbs.

1.2.2.2 Other Applications   To illustrate the versatility of LEDs in a variety of applications, this author simply looked around his home and vehicles to develop a relatively short but diverse list of applications in which LEDs are in use. Then, by going to work at a data center, where other applications made use of LEDs, additions were made to this list. The resulting list of LED applications is presented in Table 1.3. Although this list is far from being fully comprehensive, it does illustrate the wide range of application areas where LEDs are currently used. In examining the entries in Table 1.3, note that with the exception of the traffic light entry, one or more examples of the use of LEDs were provided for each application.

Although this author was able to purchase a few LED lightbulbs online at a considerable savings over current catalog prices, they are currently at a point where compact fluorescent bulbs were approximately 5 years ago. That is, approximately 5 years ago, compact

Table 1.3   General LED Application Areas

Automobiles	Household lighting	Mobile phones and PDA
Indicator lights	Bulb replacement	Backlight for screen display
Turn signal lights	Flood lights	Operational status indicator
Rear brake lighting bar	Landscape lighting	DVD player
Data Communications	Track lighting	Electronic air filter
Optical fiber transmission	LCD televisions	Television
Device indicators	Backlight	Traffic light
Displays		
Digital counter		
Digital clock		
Scoreboard		

fluorescent bulbs were relatively expensive when used as replacements for incandescent lightbulbs. However, the increasing awareness of the inefficiency of incandescent lighting, in which approximately 90 percent of the electricity used by a bulb is wasted as heat, resulted in an increase in consumer demand for alternative lighting, especially, as the cost of electricity increased. This in turn allowed manufacturers of compact fluorescent lightbulbs to expand production and reduce their retail cost, resulting in an increasing demand for the product.

In fact, in late 2007, General Electric, one of the primary manufacturers of incandescent lightbulbs, announced that it would close several manufacturing facilities, whereas it would expand its production of compact fluorescent lightbulbs. Today, LED lightbulbs cost at a minimum two to three times the cost of compact fluorescent lightbulbs. However, because they are considerably more efficient than fluorescent lighting, this author believes that within 5 years, they will replace a significant number of fluorescent bulbs in a manner similar to how compact fluorescent bulbs are currently replacing incandescent bulbs. Later in this book when we examine LED lighting, we will go into considerably more detail concerning the economics associated with LED lightbulbs as well as such issues as the expected life and heat generation of LEDs in comparison to compact fluorescent and incandescent lightbulbs.

2

FUNDAMENTALS OF LIGHT

In Chapter 1, we learned that a light-emitting diode (LED) can be considered to represent a p-n junction semiconductor. By using a variety of semiconductor materials and doping the material, it became possible to generate specific wavelengths of light when a voltage is applied to the LED. In this chapter, we will turn our attention to the fundamental properties of light to obtain a better appreciation for the operation of LEDs and the type of light they produce. Topics we will cover in this chapter include the principal properties of light, light metrics, how light is generated, and how colored light is formed. Because this book is focused on LEDs, we will obviously orient this chapter toward the light-emitting diode.

2.1 Properties of Light

We can define light as an electromagnetic radiation at a particular wavelength. Depending on the wavelength, light can be visible or invisible to the eye, infrared being a good example of invisible light.

Light has three primary properties. They include frequency or wavelength, intensity or brightness, and polarization or direction of wave oscillation.

2.1.1 Speed of Light

If you sat through a high-school physics class, you probably remember discussing the speed of light. When light travels, its speed is governed by the medium, with the speed of light in a vacuum fixed by definition to be 299,792,458 meters per second (m/s). When light travels within an optical fiber or another medium other than a vacuum, its speed is reduced.

2.1.2 Photons

Technically, light is carried or transported by photons. *Photon* is a term used by physicists to define the elementary particle responsible for electromagnetic phenomena. It is the transporter of all wavelengths of electromagnetic radiation including gamma rays, x-rays, ultraviolet light, microwaves, and radio waves.

The modern concept of the photon dates back to Albert Einstein who, during the period from 1905 through 1917, hypothesized that the photon was needed to explain the ability of matter and radiation to be in thermal equilibrium as well as the frequency dependence of light's energy. Over the years, the photon concept resulted in significant advances in physics, resulting in the development of quantum field theory and quantum mechanics as well as the invention of the laser and LED. Today, we now know that the photon interacts with matter by transferring the amount of energy (E) as follows:

$$E = h \times c/\lambda$$

where h is Planck's constant, c is the speed of light, and λ is the wavelength.

2.1.3 Planck's Constant

Planck's constant was defined by Max Planck in 1900, when he was working on the problem of how the radiation an object emits is related to its temperature. Planck developed a formula that was very close to the experimental data when he assumed that the energy of a vibrating molecule was quantized such that it could take on certain values. Because the energy would be proportional to the frequency of vibration and appeared to be represented by the frequency multiplied by a constant, Planck defined a constant, which is now known as Planck's constant in his honor. Designated by the letter "h," it has the following value:

$$h = 6.626 \times 10^{-34}\,J \cdot s$$

where J represents a Joule, named in honor of James Joule, who examined the relationship between heat and energy in the 19th century, with the Joule representing 1 kg m^2/s^2. Thus, a Joule is similar to measuring area in m² (distance times distance) or speed in m/s as the later represents distance divided by time. Returning to the afore-mentioned equation, $J \cdot s$ is a Joule second, which represents angular momentum. Based on Planck's work, Einstein later postulated that light similarly delivers its energy in a minute series of particles or quanta known now as photons, each having an energy of Planck's constant times its frequency. Both later in this chapter as well as in Chapter 3, we will discuss the orbit of electrons around an atom's nucleus and how the movement from one orbit to another results in energy that causes the emission of light in the form of a proton's wave-length. This is turn results in light being emitted at a specific color or as nonvisible light in the infrared area; the latter is used in many remote controls.

2.1.4 Frequency, Energy, and Light

Einstein's work on photoelectricity indicates the relationship between frequency, energy, and light. Thus, high-frequency photons have more energy so they make electrons flow faster. This means that light with the same intensity but a higher frequency increases the kinetic energy of the emitted electrons. Similarly, if you maintain the frequency but increase the intensity, more electrons should be emitted because more photons hit them. However, they won't come out any faster as each individual photon still has the same energy. In the opposite direc-tion, if you have lower frequency, then you will reach a point where photons will not have sufficient energy to force an electron out of an atom no matter how high the intensity is. Although probably best known for his theory of relativity, this relationship between photons, frequency, and energy formed the basis for the photoelectric effect for which Albert Einstein won the Nobel Prize.

2.1.5 Frequency and Wavelength

Returning to the equation in Section 2.1.2, which defined the amount of energy transferred by a photon, we note that whereas Planck's con-stant and the speed of light are constants, the wavelength can vary. In an LED environment, the wavelength of photons emitted from a semiconductor material determines the color of illumination. Thus, in

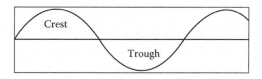

Figure 2.1   Sine wave oscillating at 1 Hz.

the following sections, we will discuss what frequency represents as well as the relationship between frequency and wavelength and, using our understanding of the latter, examine the spectrum of electromagnetic radiation including visible light and infrared that are generated by different types of LEDs.

2.1.5.1 Frequency Frequency represents the number of repeated events per unit of time. We can obtain an appreciation of frequency by examining a sine wave, which many of us may remember from our first physics class. The sine wave shown in Figure 2.1 has exactly one cycle per second. That cycle is now referred to as 1 Hertz (Hz) in honor of the German physicist Heinrich Hertz. If you double the frequency shown in Figure 2.1, the sine wave would then oscillate at a frequency of 2 Hz. Note that the point of maximum displacement (A) is called the *amplitude*, and the crest and trough of the wave are the locations of maximum and minimum displacements.

Frequency can be expressed as the reciprocal of the time or period (*T*) as follows:

$$F = 1/T$$

2.1.5.2 Frequency of Waves Frequency also has an inverse relationship with the wavelength (λ) of a wave. In this relationship, frequency (f) is equal to the speed (v) of the wave divided by its wavelength. That is

$$f = v/\lambda$$

When we discuss electromagnetic waves, v can be approximated by the speed of light (c) in a vacuum, and frequency then becomes

$$f = c/\lambda$$

2.1.5.3 The Electromagnetic Spectrum We can obtain an appreciation of where LED-generated light falls within the electromagnetic spectrum

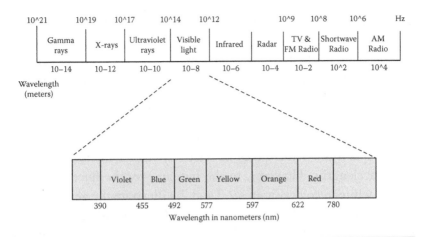

Figure 2.2 The electromagnetic spectrum and visible light.

by examining the frequencies and wavelengths of the major components of the spectrum and then focusing our attention upon visible and invisible light. The top portion of Figure 2.2 illustrates a large portion of the electromagnetic spectrum in terms of the frequency and wavelength of its major components. In the lower portion of Figure 2.2, visible light is "exploded" in terms of its wavelength in nanometers.

In Figure 2.2, first note that the frequency values at the top of the chart are approximations. Electromagnetic radiations having wavelengths from approximately 370 to 770 nm is visible to the human eye. An incandescent lightbulb or a fluorescent light radiates across the visible spectrum; however, the intensity varies in different wavelengths.

2.1.6 Spectral Power Distribution

The result of a radiation is a spectral power distribution (SPD), which shows the relative radiant power (along the y-axis) generated by the light source at each wavelength (along the x-axis) in the visible spectrum as an x-y plot. The resulting SPD plot contains all the basic physical data about light and functions as the starting point for the quantitative analysis of color. A spectrophotometer is used to measure the SPD. As we will shortly note, different types of lighting have different SPD plots.

2.1.6.1 Incandescent Light Incandescent light sources have a continuous SPD from approximately 300 to 750 nm; however, the relative power is low in the blue and green regions, whereas power is much higher in the orange and red regions of the spectrum, resulting in a "warm white" color appearance to the human eye.

2.1.6.2 Fluorescent Light SPDs for fluorescents have spikes across the visible spectrum, with spikes of relatively higher intensities at certain wavelengths. However, to the human eye, the fluorescent will appear white in color. Figures 2.3 and 2.4 illustrate the general SPDs for incandescent and fluorescent lighting. On comparing the two, we note that the SPD shown in Figure 2.4 is for a "cool white fluorescent." Other types of fluorescents, such as a "deluxe cool white," will have different SPDs.

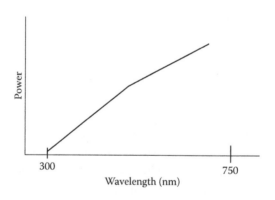

Figure 2.3 Incandescent spectral power distribution (SPD).

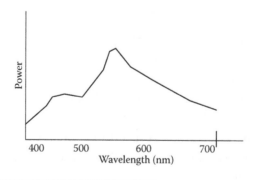

Figure 2.4 General fluorescent lighting spectral power distribution (SPD).

2.2 The CIE Color System

When discussing the properties of light, we would be remiss if we did not also discuss color. In discussing color, it's important to understand the system defined by the Commission Internationale de L'Echairage (CIE). This system can be described as an evolution of the well-known Maxwell triangle, which dates back to the 18th century and provides an easy-to-use method to define color mixing. Thus, prior to discussing the CIE system, let's first turn our attention to the Maxwell triangle.

2.2.1 The Maxwell Triangle

The Maxwell triangle was developed by the Scottish physicist James Maxwell as a mechanism to define the mixing of the three primary color components.

2.2.1.1 Overview Figure 2.5 illustrates the basic Maxwell triangle. Note that the three primary colors (green, blue, and red) are placed at the vertices of an equilateral triangle. Along the sides of a Maxwell triangle, mixing of two of the three primary color components occurs in every possible proportion. As you travel toward the center of the triangle, the third primary color becomes increasingly important, with the center of the triangle having a true white color.

If we route a line through the center from each primary color vertex to the opposite side of the triangle, we obtain the positions where cyan, yellow, and magenta occur.

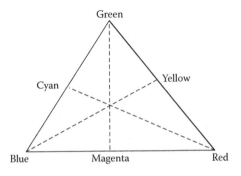

Figure 2.5   The Maxwell triangle.

2.2.1.2 Limitations Although the Maxwell triangle shows the quality aspects of colors known as *chromaticity*, which denotes hue and saturation, it does not indicate the quantity aspects of color in terms of the effective amount of light. In addition, although the triangle provides an excellent hue representation, it lacks providing a match for saturation as you compare points around the sides of the triangle. Obtaining a match of saturation levels requires the dilution of spectral color with the third primary. For example, the center point on the blue–green edge of the triangle is not as saturated as the spectral cyan. Thus, to make the two colors the same, the addition of the third primary color, red, is required. In mathematical terms, this is equivalent to adding negative red to the color in the triangle, which moves the point outside the triangle. If this process is continued for every spectral hue, the result is a curve known as the *spectral locus*. This curve shows that some of the colors reside outside the triangle.

2.2.1.3 The Spectral Locus Figure 2.6 illustrates the spectral locus curve superimposed upon Maxwell triangle. Note that because the blue–red edge of the triangle is nonspectral, it retains its straightness within the spectral locus.

The spectral locus shown in Figure 2.6 indicates that although the primary colors are the most intense obtainable, their additive mixtures cannot be used to produce the entire color spectrum. To get around this problem, the CIE used primaries that are not found in the spectrum. In doing so, the CIE selected three primaries that they called *X*, *Y*, and *Z*, which are theoretically defined as *supersaturated*

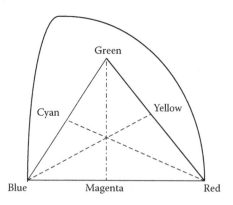

Figure 2.6 The spectral locus and Maxwell triangle.

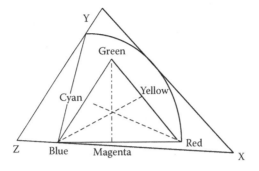

Figure 2.7 The CIE theoretical primaries and their relationship to the spectral locus.

colors; they reside outside the bounds of the spectral locus. Therefore, the CIE *X*, *Y*, *Z* system does not have to use negative values.

2.2.2 CIE Theoretical Primaries

Through the use of the three aforementioned imaginary primaries, we obtain a method that makes the coordinate system simpler. Figure 2.7 illustrates the CIE *X*, *Y*, and *Z* theoretical primaries and their relationship to the spectral locus. Note that through the use of *X*, *Y*, and *Z*, we can now define a color of wavelength λ as follows:

$$C_\lambda = xX + yY = zZ$$

where *x*, *y*, and *z* are the amounts of the primaries *X*, *Y*, and *Z*. These three values represent relative proportions that can be normalized to obtain the values of *X*, *Y*, and *Z* as follows:

$$x = X/(x + y + z), y = Y/(x + y + z), z = Z/(x + y + z)$$

2.2.3 CIE Chromaticity Chart

Because the values for *x*, *y*, and *z* were normalized, $x + y + z = 1$ and we now require only two coordinates to compute the third. The values of *x* and *y* are known as *chromaticity coordinates*, as they are limited to containing hue and saturation information. However, the *y* value was modified to carry luminosity, so that any color can now be represented by using its chromaticity values *x* and *z* and its *y* value, resulting in the CIE chromaticity chart, which is shown in Figure 2.8.

The CIE chromaticity chart shown in the figure characterizes colors by a luminance parameter (Y) and two color coordinates (X and Y)

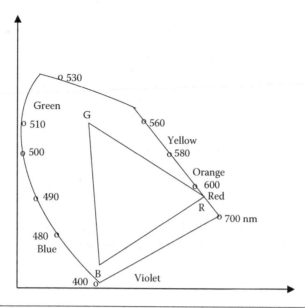

Figure 2.8 The CIE chromaticity chart, showing wavelengths marked around the outside.

which specify a point on the chart. The original CIE chart dates back to 1931, with revisions made in 1960 and 1976. The 1931 version of the chart remains the most widely used version and enables colors to be matched by combining a given set of three primary colors represented on the chart by a triangle joining the coordinates of the three colors.

The x- and y-axis values shown in Figure 2.8 represent the mapping of human color perception in terms of the CIE x and y parameters. The gamut of human color vision can be roughly described on the CIE chart shown in Figure 2.8; however, doing so would be an approximation, and this author suggests that the use of wavelengths will provide a better indication of color. Thus, as we return to a discussion of LED light, we will also examine the wavelength of LED colors.

2.3 LED Light

Unlike incandescent and fluorescent lights, LEDs emit near-monochromatic light. That is, an LED emits light at a specific wavelength; this explains why this type of diode is efficient for colored light applications.

2.3.1 *Comparing LEDs*

Table 2.1 contains an LED color chart, which indicates the wavelength, color, voltage and current rating, and doping material for a series of LEDs currently available from different vendors. Note that the information in this table is not all inclusive, and it is highly likely that there are other LEDs that vary from the properties shown. However, by examining the entries in the table, you can obtain at a minimum the ability to associate LED color and wavelength, as well as an appreciation for the variance in forward voltage (V_f), and knowledge about the doping materials used to manufacture LEDs.

In the entries listed in Table 2.1, note that the first three LEDs emit invisible or infrared light. These LEDs are commonly used in communications and other applications that require the use of an infrared light source, such as remote controls for TVs, DVD players, and similar electronic equipment.

2.3.2 *White Light Creation Using LEDs*

It should be noted that there are currently several methods employing LEDs that are commonly used to generate white light. One method of making white light with LEDs is by mixing monochromatic LEDs. A second method is coating certain types of LEDs with a phosphor.

In this section we will initially review the primary methods used to create white light using LEDs. In Section 2.3.8 we will go into considerable more detail concerning the generation of white light by LEDs.

2.3.2.1 White Light Creation by Mixing Colors   Red, green, and blue, and occasionally amber, light from monochromatic LEDs is mixed to generate white light. The result is referred to as an *RGB-generated white light*.

2.3.2.2 White Light Creation Using Phosphor   A phosphor-coated LED or a blue or near-ultraviolet LED is also used to produce white light. Because they are based on a blue or near-ultraviolet LED, they usually have very high color temperatures, which results in a cool or blue

Table 2.1 LED Color Chart and the Basic Device Properties

WAVELENGTH (MM)	COLOR NAME	FORWARD VOLTAGE @ 20 MA	INTENSITY 5 MM: LED	VIEWING ANGLE (DEGREES)	LED DYE MATERIAL
940	Infrared	1.5	16 mW@50mA	15	GaAlAs/GaAs
880	Infrared	1.7	18 mW @ 50 mA	15	GaAlAs/GaAs
850	Infrared	1.7	26 mW @ 50 mA	15	GaAlAs/GaAs
660	Ultra red	1.8	2000 mcd @ 50 mA	15	GaAlAs/GaAs
635	High efficiency red	2.0	200 mcd @ 20 mA	15	GaAsP/GaP
633	Super red	2.2	3500 mcd @ 20 mA	15	InGaAlP
623	Red-orange	2.2	4500 mcd @20 mA	15	InGaAlP
620	Super orange	2.2	4500 mcd @ 20 mA	15	InGaAlP
612	Orange	2.1	160 mcd @ 20 mA	15	GaAsP/GaP
595	Super yellow	2.2	5500 mcd @ 20 mA	15	InGaAlP
592	Amber yellow	2.1	7000 mcd @ 20 mA	15	InGaAlP
585	Yellow	2.1	100 mcd @ 20 mA	15	GaAsP/GaP
3500 K	"Incandescent" white	3.6	200 mcd @ 20 mA	20	SiC/GaN
4500 K	"Incandescent" white	3.6	2000 mcd@20 mA	20	SiC/GaN

5000 K	Pale white	3.6	4000 mcd@20 mA	20	SiC/GaN
6500 K	Pale white	3.6	4000 mcg/20 mA	20	SiC/GaN
8000 K	Cool white	3.6	6000 mcd @20 mA	20	SiC/GaN
574	Super lime yellow	2.4	1000 mcd @ 20 mA	15	InGaAlP
570	Super lime green	2.0	1000 mcd @ 20 mA	15	InGaAlP
565	High efficiency green	2.1	200 mcd @ 20 mA	15	GaP/GaP
560	Super pure green	2.1	350 mcd @ 20 mA	15	InGaAlP
555	Pure green	2.1	80 mcd @ 20 mA	15	GaP/GaP
525	Aqua green	3.5	10,000 mcd @ 20 mA	15	SiC/GaN
505	Blue green	3.5	2000 mcd @ 20 mA	15	SiC/GaN
470	Super blue	3.6	3000 mcd @ 20 mA	15	SiC/GaN
430	Ultra blue	3.8	100 mcd @ 20 mA	15	SiC/GaN

Note: GaAlAs/GaAs–Gallium aluminum arsenide/gallium arsenide.
GaAsP/GaP–Gallium arsenic phosphide/gallium phosphide.
GaP/GaP–Gallium phosphide/gallium phosphide.
InGaAlP–Indium gallium aluminum phosphide.
SiC/GaN–Silicon carbide/gallium nitride.

appearance. By adding phosphors that emit in the red area of the visible spectrum, it becomes possible to obtain a warmer white color, although the luminous efficacy of the LED is reduced by approximately 50 percent. Later in Section 2.3.8 we will cover in more detail the primary methods used to emit white light by using an LED as the illumination source.

2.3.3 *Intensity of an LED*

One of the columns in Table 2.1 labeled "Intensity" deserves a degree of explanation. The intensity of an LED can be defined in terms of millicandela (mcd), which represents 1/1000 of a candela. To obtain an appreciation for what the candela represents, a brief overview of an older term known as *candlepower* is in order.

2.3.3.1 Candlepower Candlepower, which is abbreviated as cp, is an obsolete but still well-known term used as a measurement for luminosity based on the light emitted by a candle manufactured to a specific formula. The term candlepower dates back to 1860 when it was defined in England by the Metropolitan Gas Act as the light emitted by a pure spermaceti candle weighing one-sixth of a pound, which burned at the rate of 120 g/hr. Note that spermaceti is obtained from the head of a whale and at one time was used to create light quality candles. As you might expect, across the English Channel, the French standard of light differed from the English method, whereas in Germany a third definition was used. This divergence in the definition of candlepower resulted in several international meetings. In 1927, the international candle was redefined in terms of a carbon filament incandescent lamp, whereas in 1937, the international candle was again redefined, this time in terms of the luminous intensity of a blackbody at the freezing point of liquid platinum which was set at 58.9 international candles per square centimeter. In 1948, the use of the term candlepower was replaced by the international unit known as *candela* (cd), with the candlepower being approximately 0.981 cd; but in the modern literature, the term candlepower is synonymous with the term candela.

2.3.3.2 The Candela   A candela represents a base unit of luminous intensity, which is defined in the Merriam-Webster online dictionary as "being equal to the luminous intensity in a given direction of a source which emits monochromatic radiation of frequency 540 x 10^{12} hertz and has a radiant intensity in that direction of watt per unit solid angle."

This definition of the candela represents a luminous intensity in a given direction at a precise frequency. In addition, as we will note later in this chapter, the candela is also defined as having a radiant intensity of 1/683 watts per steradian (W/sr).

The frequency selected for the definition of the candela corresponds to a wavelength of approximately 555 nm, which represents the visible spectrum of light near the color green. The selection of this point was based on the fact that the human eye is most sensitive to the selected frequency when adapted for bright conditions. At other frequencies, additional radiant intensity is needed to obtain the same level of luminous intensity when viewed via the human eye. The relationship between the luminous intensity in candelas $I_v(\lambda)$ of a particular wavelength (λ) is given by the following formula:

$$I_v(\lambda) = 683.00 \, y(\lambda) \, I(\lambda)$$

where

$I_v(\lambda)$ = the luminous intensity in candelas

$I(\lambda)$ = the radiant intensity in watts per steradian

$y(\lambda)$ = the standard luminosity function if more than one wavelength is present, which then requires the sum over the spectrum of wavelengths present to obtain the total luminous intensity

Thus, $I_v = 683 \left(\sum\limits_{0}^{\infty} \right) I(\lambda) \, y(\lambda) \, \Delta\lambda$

2.3.4 On-Axis Measurement

Because the intensity of an LED depends on the on-axis measurement, it's possible for a 100 mcd LED to produce less light than an 80 mcd device. This results from the fact that the millicandela rating is determined by an on-axis measurement of peak intensity at a specific current and not by the measurement of total light output. Thus, a diffused LED that spreads light over a wide viewing angle could have an on-axis intensity of 80 mcd yet emit more light than a nondiffused LED whose on-axis intensity is 100 mcd or more.

In Table 2.1, both the luminous intensity and viewing angle are provided and need to be considered when determining total light output. That is, when two LEDs have the same luminous intensity value, the device with the larger viewing angle will always have the higher total light output, which explains the prior example (in which an 80 mcd LED can have a greater light output than a 100 mcd LED).

2.3.5 Theta One-Half Point

It's important to note that because of its shape, the LED encapsulation functions as a lens that will magnify light emitted from the LED. The off-axis point where the intensity of the LED is half its on-axis intensity is referred to as *theta one-half* (q ½). Thus, twice the q ½ value represents the LED's full viewing angle.

Figure 2.9 illustrates the on-axis luminous intensity value (I_v) with respect to its theta one-half point. Note that the intensity is normally obtained through the use of a photometer and that light is visible beyond the q ½ point.

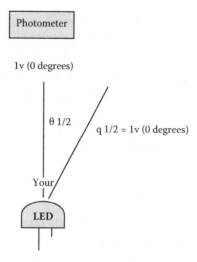

Figure 2.9 Measuring luminous intensity and the theta one-half point of an LED.

2.3.6 Current and Voltage Considerations

The luminous intensity of an LED is approximately proportional to the amount of current supplied to the device. Thus, the greater the current, the higher is the intensity. However, design limits will result in an upper boundary on both current and light intensity.

Most LEDs are designed to operate at a current of 20 mA. When considering an LED for a specific application, you need to consider the operating current of the LED in comparison to the amount of heat that can be tolerated by the application also. For example, LEDs that are designed to operate at 12 V would have a greater heat dissipation than LEDs designed to operate at 6 V. Because heat dissipation is a significant factor in the life of an LED, it's important to consider the current rating of the LED and its applied and forward voltage if your application requires a long-life LED.

In addition, it's also important to consider the location of the LED as this will govern its ability to dissipate heat. For example, although most LEDs used as indicators in monitors, televisions, and various types of toys do not reside under a plastic panel, in certain instances products are designed to include this type of panel as a protective measure against glare, dirt, dust, or other elements. In this situation, the plastic panel will adversely affect the ability of the LED to dissipate the generated heat.

2.3.7 Lumens, Candelas, Millicandelas, and Other Terms

When comparing LEDs with incandescent lighting, it's important to note that a direct comparison of the two is not possible. This issue results from the fact that the amount of light emitted from an LED is specified by the measurement at a single point, known as the *on-axis luminous intensity value* (I_v). This measurement of an LED's luminous intensity is not directly compatible with the light produced by incandescent lightbulbs, with the latter using spherical candlepower as a mechanism to denote the quantity of light emitted by the lightbulb. Thus, to obtain a solid appreciation for the relationship between lumens and candela, let's turn our attention to the manner in which these units are related.

2.3.7.1 Lumens A lumen represents a unit of light output, which is equal to 1/60 × π of the light emitted by one square centimeter of an ideal blackbody surface at the melting point of platinum, with π equal to the constant 2.14159 …. Lumens were defined in terms of the 1931 official photopic function that was modified in 1988, which essentially examines the quantity of light produced at every wavelength that is present. One watt of light at any single wavelength represents a lumen (unit is lm), which is mathematically 681 times the official photopic function of the wavelength.

2.3.7.1.1 Watts as a Measurement Tool To alleviate some potential confusion concerning watt numbers, let's differentiate between the watt number used for light output and the unit used for measuring electrical power. When using the term watt for measuring electrical power, it represents volt times the amperage (ampere) in a direct current (dc) circuit, that is,

$$watt = volt \times ampere$$

If we are referring to a single-phase ac circuit, then

$$watt = volt \times ampere \times power\ factor$$

where power factor represents the ratio of real power to apparent power and is a number between 0 and 1. In comparison, the watt number used to express light output, as previously noted, is mathematically 681 times the official photopic function of the wavelength. To simplify this, manufacturers use the watt number as a mechanism to compare light output to an incandescent lightbulb output. Then, a 60 W light output is equal to a 60 W incandescent lightbulb. To further muddy the water, often two watt numbers are used in a product specification, one for the electrical power consumption whereas the other is used to specify light output. Fortunately, reading the fine print on the package usually alleviates the potential confusion.

Returning our attention to the standard 100 W lightbulb, we can obtain knowledge of different lighting terms. The standard 100 W, 120 V incandescent lightbulb is typically rated for 750 hr and emits 1710 lm. The lumens rating represents the unit of luminous flux and is equal to the light radiated by a source of 1 cd intensity radiating equally in all directions. Thus, in some ways, we can actually think of a 1710 lm incandescent lightbulb as providing the light of 1710 candles, because a candela can be thought of being loosely equivalent to a burning candle. Lumens are used to measure and compare lightbulbs as stand-alone light sources. You can usually examine the packaging of a lightbulb and obtain its rated life in terms of hours of operation, its wattage, its voltage, and amount of light it provides in lumens.

2.3.7.2 Lumens per Watt and Lux   Two terms that require our attention are *lumens per watt* and *lux*. The term lumens per watt provides an indication of the efficiency of a lightbulb, including its efficiency in converting electrical energy into light. In comparison, lux is used to measure the light intensity produced by a lighting fixture and is measured in lumens per square meter. Thus, the higher the lux reading, the more light the fixture provides over a given area.

2.3.7.3 Watt Dissipation   If you examine the packaging on an LED, you can usually find information concerning the number of watts the device will dissipate. If this information is not displayed on the package, you can easily compute the value by determining the forward voltage (V_f) and milliamp (mA) rating of the LED. In some cases, both V_f and mA are listed on the package, with watts dissipated curiously missing. Thus, let's review how we can compute the watts dissipated by an LED.

Let's assume you plan to use an orange LED that has a forward voltage rating of 2 V and a current rating of 20 mA. Because wattage is obtained by multiplying voltage by current ($W = I \times V$),

$$\text{watts} = 2 \text{ V} \times 0.20 \text{ A} = 0.04$$

When you carefully examine certain types of LED packaging, you may note that its emission in lumens is also provided. On other LED packages, you may find the metric lumens/watt cited, which actually represents the efficiency of the LED. The latter provides a mechanism to compare the output efficiency of different LEDs with one another as well as with various types of other light sources.

If the LED package indicates efficiency in terms of lumens per watt, you can compute the watts dissipated or use the value from the package and multiply it by the lumens per watt value. Doing so will allow you to obtain the LED emission in lumens. Thus, there are several minor computations you can perform to determine an LED's efficiency and heat dissipation, if those metrics are not provided on its packaging.

2.3.7.4 Steradian Previously in this chapter, we noted the dictionary definition of a candela in terms of radiation frequency and direction on an on-axis measurement (see Section 2.3.3.2). In doing so, we avoided defining the term steradian until now.

The term steradian is important as it provides an alternate mechanism for defining a candela. That is, we can define a candela as one lumen per steradian; thus, we need to know what a steradian represents to better understand this definition.

A steradian (sr) is a unit of solid-angle measure in the International System of Units (ISU). It is defined as the solid angle of a sphere subtended by a portion of the surface whose area is equal to the square of the sphere's radius. Because the complete surface area of a sphere is 4π times the square of its radius, this results in the total solid angle about a point being equal to 4π sr. Thus, the steradian is 1(4 × π) of a whole sphere or 1/(2 × π) of a hemisphere, which is approximately 3283 "square degrees." If you know the beam angle, you can compute steradians using the following formula:

$$\text{steradians} = 2 \times \pi \times (1 - \cos(0.5 \times (\text{beam angle})))$$

Based on the preceding equation, you can obtain a rough approximation for the light output of an LED in lumens by first determining the steradian beam coverage. Next, you have to multiply the steradian beam coverage by the candela value expressed either in terms of candela or millicandela.

That is, because

$$\text{candela} = \text{lumens/steradian beam coverage}$$

then

$$\text{lumens} = \text{candela} \times \text{steradian beam coverage}$$

or because 1000 mcd equals 1 cd, we can express lumens in terms of millicandela and steradian beam coverage as follows:

$$\frac{\text{lumens}}{1000} = \text{millicandela} \times \text{steradian beam coverage}$$

Table 2.2 Converting Lumens to Millicandelas

RADIATION ANGLE (DEGREES)	DIVIDE BY
5	167.22
10	41.82
15	18.50
20	10.48
25	6.71
30	4.67
35	3.44
40	2.64
45	2.09

2.3.7.4.1 Converting Millicandela to Lumens Due to the relationship (noted in Section 2.3.7.4) between lumens and millicandela, it becomes quite easy to determine one metric if you know the other metric and the coverage beam, which is also commonly referred to as the *radiation angle*. To assist readers in this computation, Table 2.2 indicates the divisor to be used based on beam angles varying from 5 to 45° when computing an LED's luminous intensity in luminous flux and luminous intensity millicandela given its luminous flux in lumen.

Note that we just learned two new terms. That is, luminous flux represents a measure of the quantity of energy that is transmitted by a light source in all directions and is denoted by lumen (lm). In comparison, the luminous intensity of a light source is the density of luminous flux emitted in a given direction and is expressed in millicandela (mcd).

In addition to the candela, steradian, luminous flux, and luminous intensity, there are several additional terms commonly used to describe the properties of light. In the following sections, we will briefly become acquainted with these terms as it is highly likely that readers will encounter them; this information should assist in expanding our knowledge of lighting terminology.

2.3.7.5 Luminous Energy Another term that you may see on occasion, which is often misleading, is *luminous energy*; many persons confuse it with luminous intensity. Recall that luminous intensity of a light source represents the density of luminous flux emitted in a given direction. In comparison, luminous energy is a term used to denote the perceived energy of light. Because the human eye is limited to seeing light in the visible spectrum, luminous energy will considerably differ from luminous intensity in the infrared spectrum and more than likely differ in value from person to person in the visible spectrum due to differences in the manner in which humans see light.

2.3.7.6 Illuminance   *Illuminance* represents the total luminous flux incident on a service and is measured in lumens per square meter (lm/m^2). Because illuminance represents a measure of light at a point in time, it is also referred to as *brightness*. In some premium cable channel movies that have a "before the scene" show, it's quite common to see a studio technician use a lux meter to check the illuminance level for adjusting video cameras prior to the shout of "action."

2.3.7.7 Lighting Efficiency   *Luminous efficiency*, as we will shortly note, is also referred to as *efficiency*. This term enables comparison of the efficiency of different types of lighting.

Luminous efficiency is measured in lumens/watt (lm/w) and provides a measurement of the efficiency of a light emitter when its output is adjusted to account for the spectral response curve. When expressed as a dimensionless value, the overall luminous efficiency is referred to as the *lighting efficiency*. Table 2.3 provides examples of the overall luminous efficiency and luminous flux for eight light sources. Note

Table 2.3   Luminous Efficiency and Efficiency Examples

	OVERALL LUMINOUS EFFICIENCY (LM/W)	OVERALL LUMINOUS EFFICIENCY (PERCENTAGE)
Incandescent		
5 W tungsten	5	0.7
40 W tungsten	12.6	1.9
100 W tungsten	17.5	2.6
Fluorescent		
5–24 W compact fluorescent	45–60	6.6–8.8
34 W tube	50	7.0
Halogen		
Glass	16	2.3
Quartz	24	3.5
LED		
White	20–70	3.8–10.2

that luminous efficiency represents the ratio of emitted luminous flux to radiant flux, whereas overall luminous efficiency indicates efficiency of the conversion of electrical energy to optical power.

2.3.7.8 Color Temperature   Color temperature provides a measure of the color of a light source that is relative to a black body at a particular temperature. This temperature is expressed in degrees Kelvin (K), where Kelvin represents a thermodynamic temperature scale in which the coldest temperature possible is zero Kelvin (0 K). The Kelvin scale and the unit Kelvin are named after the physicist and engineer William Thomson (first Baron Kelvin), who identified the need for an absolute thermometric scale. Absolute zero (0 K) is equivalent to −459.63° F and −273.15° C. Another term related to Kelvin that you will occasionally encounter is *mired*. Mired represents the color temperature in Kelvin divided by one million.

2.3.7.9 Representative Lighting Color Temperature   Incandescent lights have a low color temperature of approximately 2800 K. In comparison, daylight has a high color temperature at or above 5000 K. In between the two are popular fluorescent lighting referred to as "cool white" at approximately 4000 K and "white" or "bright white" at approximately 2900–3100 K. Table 2.4 provides a summary of a variety of lighting and their color temperatures or temperature range in Kelvin.

In general, lighting having a low color temperature at approximately 2200 K has a red-orange tone. At 2800 K, lighting has a red-yellowish tone and the color is typically referred to as "warm white" or "soft white." Daylight has a relatively high color temperature at or above 5000 K and appears bluish. In between, a halogen lamp at 3000 K generates a yellowish light, whereas the color of the most popular fluorescent light, which is rated at approximately 4000 K, is referred to as "cool white." At the high end of the color temperature, an LED "cool white" occurs at a color temperature of 8000 K. Readers are referred to Table 2.1 for a list of common LEDs and their color temperatures.

Table 2.4 Representative Color Temperatures for a Variety of Lighting

TYPES OF LIGHTING	KELVIN
High-pressure sodium	2200
Incandescent (soft white)	2800
Halogen	3000
Fluorescent (cool white)	4000
Daylight	>5000
LED (cool white)	8000

2.3.8 LED White Light Creation

Initially, the first series of LEDs emitted red light, followed by yellow and orange. To emit white light, the LED manufacturer commonly uses one of three approaches: wavelength conversion, color mixing, or a technology referred to as *homoepitaxial ZnSe*.

2.3.8.1 Wavelength Conversion Wavelength conversion involves converting all or a part of an LED's emission into visible wavelengths that are perceived as white light. Currently, there are several methods that have been developed over the years to generate white light using LEDs. Some of these methods include the use of blue LED and yellow phosphor; blue LED and several phosphors; ultraviolet LED and blue, green, and red phosphors; and an LED with quantum dots.

2.3.8.1.1 Blue LED and Yellow Phosphor In this method of wavelength conversion, blue light from an LED is used to excite a yellow phosphor, resulting in the emission of yellow light. The resulting mixture of blue and yellow light results in the appearance of white light. This method of wavelength conversion is considered to be the least expensive method for producing white light.

2.3.8.1.2 Blue LED and Several Phosphors In this method of wavelength conversion, the use of multiple phosphors results in each phosphor emitting a different color. These emissions are combined with the original blue light to produce white light.

Compared to the use of a blue LED and a yellow phosphor, the use of multiple phosphors results in white light having a broader wavelength spectrum and a higher color quality. However, the use of multiple phosphors makes this method more expensive than the former.

2.3.8.1.3 Ultraviolet LED with RGB Phosphors A third wavelength conversion method involves the use of an ultraviolet LED with red, green, and blue (RGB) phosphors. Ultraviolet light is used to excite the red, green, and blue phosphors, whose emissions are mixed to provide a white light having a broad wavelength and rich spectrum.

2.3.8.1.4 Blue LED and Quantum Dots   This method involves the use of a blue LED and quantum dots. Quantum dots are extremely small semiconductor crystals that can be between 2 and 10 nm, which corresponds to 10–50 atoms in diameter. When used with a blue LED, the quantum dots represent a thin layer of nanocrystal particles that contain 33 or 34 pairs of cadmium or selenium that are coated on top of the LED. The blue light emitted by the LED excites the quantum dots. This action results in the generation of a white light that has a wavelength spectrum similar to the ultraviolet LED that uses RGB phosphors.

2.3.8.2 Color Mixing   This method for generating white light involves using multiple LEDs in a lamp and varying the intensity of each LED. Referred to as *color mixing*, a minimum of two LEDs are used, generating blue and yellow emissions that are varied in intensity to generate white light. Color mixing can also occur using three LEDs, where red, blue, and green are mixed, or four LEDs where red, blue, green, and yellow are mixed. Because phosphors are not used in color mixing, there is no loss of energy during the conversion process; as a result, color mixing is more efficient than wavelength conversion.

2.3.8.3 Homoepitaxial ZnSe   A third method for generating white light is based on a technology referred to as homoepitaxial ZnSe. This technology was developed by Sumitomo Electric Industries, Ltd., Osaka, Japan, which teamed with Procomp Informatics, Ltd., Taipei, Taiwan, to commercialize the technology under a joint venture that was named Supra Opto, Inc.

A homoepitaxial-ZnSe-based white LED is produced by growing a blue LED on a zinc selenide (ZnSe) substrate, which results in the simultaneous emission of blue light from the active region and yellow from the substrate. Because no phosphors are used, this approach makes packaging less complicated and increases the overall efficiency of the device. In addition, the elimination of phosphors eliminates any potential patent problems.

In a research study, it was found that the epitaxial layer of the LED emitted a greenish blue light at a wavelength of 483 nm, whereas the ZnSe substrate simultaneously emitted an orange-colored light at a wavelength of 595 nm. Together, this results in a white LED whose operating characteristics include an operating voltage of 2.7 V, a 20 mA current, an optical output of 20 mW, a luminous efficiency of 8 lm/W, minimum lifetime of 8000 hr until the optical output decreases to half the initial value, and a color temperature of white light in the range of 3000 K and above. Currently, this LED is being used in a range of applications such as lighting, indicators, and backlights for liquid crystal displays. As its life is expected to improve due to further development efforts, this LED will become suitable for additional applications.

3

LEDs Examined

In the previous chapters of this book, we learned what we can consider as basic information concerning light-emitting diodes (LEDs), which provides us with a solid foundation for understanding how LEDs work. This information included the basics of p-n junction and emission of both infrared and visible light. In this chapter, we will probe deeper into the manner in which LEDs operate, focusing our attention directly on the operation of a p-n junction. In addition, we will cover two additional LED topics in this chapter that will serve to expand our knowledge of LEDs. The first topic concerns a relatively new type of LED referred to as an *organic LED*. The organic LED has the potential to make science fiction a reality, with the possibility of having a large-screen super-thin TV that can be rolled up and put away when not in use. The second topic that will be discussed in this chapter concerns the use of an LED driver. Thus, sit back and relax as we continue our exploration of the light-emitting diode.

3.1 P-N Junction Operation

The LED is a special type of diode, which is a very simple semiconductor device. The diode is a two-terminal electronic device that enables electric current to pass primarily in one direction, with the current dependent on the voltage between the leads.

When we consider the evolution of the LED from a diode, we see that the key to an LED's ability to emit light is its p-n junction and the doping of a substrate material with different materials to form the junction. As we will shortly note in detail, the dopant in the n-region provides mobile negative charge carriers known as *electrons*, whereas the dopant in the p-region provides mobile positive charge carriers referred to as *holes*. Thus, when a forward voltage is applied to the p-n junction from the p- to the n-region, the charge carriers are injected across the junction into a zone where they recombine and convert their excess energy into light.

3.1.1 Semiconductor Material

There are two main types of semiconductor materials: *intrinsic* and *extrinsic*. In intrinsic semiconductor materials, semiconducting properties occur naturally. On the other hand, in extrinsic semiconductor materials, semiconductor properties are the result of external processes. Today, almost all semiconductors are extrinsic, as this allows the properties of the material to be explicitly defined. To do so, the material is doped by the addition of "foreign" atoms.

LEDs are solid-state devices. As such, one of the critical elements that govern the operation of the device is its p-n junction. As we noted in Chapter 1 of this book, when p-type and n-type materials are placed in contact, the resulting junction behaves very differently from either type of material by itself. That is, current will flow in one direction, referred to as *forward bias*, but not in the reverse direction, referred to as *reverse bias*. The inability to reverse bias results from the charge transport process in the p-type and n-type silicon used to form the p-n junction.

3.1.2 Basic Concepts of Atoms

Recall that the core of an atom is its nucleus. The nucleus contains one or more protons and may contain one or more neutrons. Protons are positively charged, whereas neutrons have no charge. Orbiting around the atom are one or more electrons. Electrons are relatively small in comparison to protons and neutrons; however, they have a negative charge.

As an example of the relationship between electrons, protons, and neutrons, let's consider the simplest of all atoms, the hydrogen atom. As illustrated in Figure 3.1, the hydrogen atom has one proton and one electron; however, the proton has approximately 1850 times the mass of its electron. Elements are classified by the number of protons they have, which becomes their atomic number. Thus, hydrogen has an atomic number of 1, whereas helium that has two protons has an atomic number of 2, and so on.

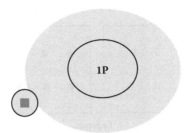

The hydrogen atom has one proton and one electron

Figure 3.1 The hydrogen atom.

3.1.2.1 Electrical Charge An atom in its initial unaltered state has the same number of electrons as it has protons. At this point of time, the atom's total electrical charge is said to be balanced. If the atom loses an electron, it will have more protons than electrons and its total charge will become positive. Conversely, if the atom gains an extra electron, it will have more electrons than protons and it will have a negative electrical charge. In physics, an atom with a positive charge is referred to as a *positive ion*, whereas an atom with a negative charge is referred to as a *negative ion*. A positive ion will try to attract an electron (positives attract negatives), whereas a negative ion (repel negatives) will attempt to lose its extra electron.

3.1.2.2 Band Theory Electrons orbit around the nucleus in energy levels or bands. As the number of electrons increases, they fill the bands in a predefined order. That is, the innermost band is filled first, followed by the population of band two, and so on. Band one can hold two electrons, whereas band two can hold eight, and the last band that holds electrons in an atom is referred to as the *valence band*. The first unfilled level above the valence band is referred to as the *conductor band*. As the bands fall further out from the nucleus, their energy level increases.

3.1.3 Energy Bands

Figure 3.2 illustrates the energy bands in the p-type and n-type silicon at equilibrium. The open circles in p-type silicon on the left side of the junction shown in Figure 3.2 represent holes or deficiencies

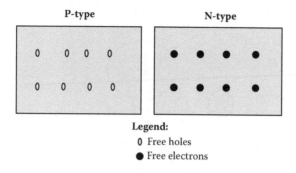

Legend:
0 Free holes
● Free electrons

Figure 3.2 Energy bands in the p-type and n-type silicon at a point of equilibrium.

of electrons in the semiconductor material, which can act as positive charge carriers. In comparison, the n-type material contains free electrons. A p-n junction is created when you bring p-type and n-type materials together to make one piece of semiconductor material. The free electrons on the n-side and the free holes on the p-side will initially wander across the junction. To obtain a better understanding of the activity at the p-n junction, let's focus our attention on the atom and its components.

3.1.4 Conduction and Valence Bands of Conductors, Semiconductors, and Insulators

Figure 3.3 illustrates the general relationship between conduction and valence bands for conductors, semiconductors, and insulators. Note that electrons in the valence band do not participate in the conduction process. To participate in the conduction process, electrons have to be in the conduction band. Thus, the further the electron is from the nucleus of the atom and the fewer neighbors an electron has in the

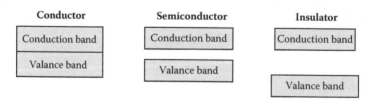

Figure 3.3 Relationship between the conduction and valence bands of conductors, semiconductors, and insulators.

valence band, the smaller the gap between the valence and conduction bands. As the gap decreases, less energy is required to move an electron to the conduction band where it becomes a free electron and can then participate in the conduction process.

Returning our attention to Figure 3.3, you will note that the gap between the conduction and valence bands widen for semiconductors and become even wider for insulators.

In a pure metal, the gap between the conduction band and the valence band is nonexistent, and the bands actually overlap, allowing the valence electrons to easily flow to either band. This results in metals being excellent conductors. In comparison, insulators have an energy gap far greater than the energy level of the electron. Because all the electrons reside in the valence band, they cannot rise through the gap when a potential difference is applied unless an extremely high voltage is applied that enables the electrons to jump the gap into the conduction band, a process referred to as *insulation breakdown*. In between conductors and insulators are semiconductors. Semiconductors have a narrow gap between the conduction and valence bands, and through doping, which is the process of adding foreign atoms to the material, electrons obtain the ability to flow in one direction.

3.1.5 Equilibrium

Figure 3.4 shows the p-n junction energy bands at equilibrium. In this illustration, the open circles on the left side of the junction represent holes or deficiencies of electrons. Such deficiencies can act as positive charge carriers. In comparison, the solid circles to the right of the p-n junction represent available electrons from the n-type dopant. As

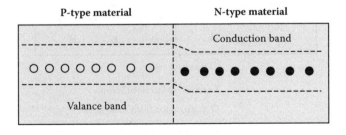

Figure 3.4 P-n junction energy bands at equilibrium.

electrons get closer to the junction, they combine with the holes, form-ing a *depletion region*. Now, let's turn our attention to this region.

3.1.5.1 Depletion Region Operation During the formation of a p-n junction, a portion of the free electrons in the n-region will diffuse across the junction. As they do so, they will combine with holes to form negative ions (atoms with negative charges), leaving positive ions (atoms with positive charges) at the donor impurity sites. Figure 3.5 illustrates the formation of negative and positive ions in the depletion region. Note that in the p-type region, the combining of electrons and holes results in the formation of negative ions, whereas in the n-type region, positive ions result from the removal of electrons.

As indicated in Figure 3.5, filling a hole results in the creation of a negative ion, leaving behind a positive ion on the n-side of the p-n junction. This action results in the buildup of a space charge that cre-ates the depletion region. This region inhibits any additional electron transfer unless a forward bias is placed on the p-n junction.

To remove the depletion zone, electrons need to move from the n-type area to the p-type area, whereas holes need to move in the oppo-site direction. This action is accomplished by connecting the n-type side of the diode to the negative voltage of a circuit and the p-type side to the positive voltage. This results in the free electrons in the n-type material being repelled by the negative charge and drawn toward the positive electrode, whereas the holes in the p-type material move in the opposite direction. When the voltage difference between the posi-tive and negative electrodes becomes sufficiently high, the electrons

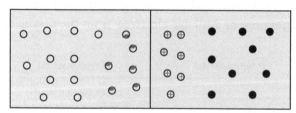

Depletion region

Legend: ● Electron ○ Hole
◔ Negative ion from filled hole
⊕ Positive ion from removed electron

Figure 3.5 Formation of negative and positive ions.

in the depletion zone will be boosted out of their holes and they start moving. Then, the depletion zone will disappear and a current will flow across the p-n junction.

3.1.5.2 Bias Effect   There are three situations that govern the bias effect on electrons in the depletion zone. They are equilibrium state, reverse bias state, and forward bias state.

3.1.5.2.1 Equilibrium State   First, the Coulomb force from ions will prevent further migration of electrons across the p-n junction. Recall that Coulomb's law states that the electrostatic force between two charged bodies is proportional to the amount of charge on the bodies divided by the square of the distance between them. This means that if the bodies are oppositely charged, they will be attracted toward one another. Conversely, if the two bodies are similarly charged (both positive or both negative) the force between them becomes repulsive. It's important to note that Coulomb's law is applicable when the charged bodies are much smaller than the distance separating the bodies. When this situation occurs, the bodies can be treated as point charges.

Coulomb's law and the unit of electrical charge, coulomb, were named in honor of Charles Augustin de Coulomb, a French physicist who is known for his research on electricity, magnetism, and friction. Coulomb's law was deducted in 1785 from experimental measurements of the forces between charged bodies made by Charles Augustin de Coulomb, which resulted in the definition of the coulomb as the amount of electrical charge transported by a current of 1 ampere (A) in 1 second (s). That is,

$$1 C = 1 A \times 1 s$$

At the equilibrium state, the migration of electrons across the p-n junction is prevented, owing to repulsion by the negative ions in the p-region and attraction by the positive ions in the n-region.

3.1.5.2.2 Reverse Bias   The second situation that governs the bias effect on electrons in the depletion region is the application of a reverse voltage, more formerly referred to as a reverse bias. When this situation occurs, electrons are driven away from the p-n junction. Because electrons from the n-region must move to the junction and combine with holes from the p-region to conduct electricity, a reverse bias prevents conduction.

3.1.5.2.3 Forward Bias The third situation that governs the bias effect on electrons in the depletion region is the application of a voltage in the forward direction, referred to as a forward bias. The forward voltage enhances the ability of electrons to overcome the Coulomb barrier of the space charge in the depletion region. In effect, this results in electrons flowing with minimal resistance in the forward direction.

The result of the previously described bias effects is a forward current that corresponds to the forward bias based on the voltage applied, the type of the semiconductor material, and the level of doping of the material. We will shortly illustrate the bias effect in the form of a chart. However, prior to doing so, a few words are in order concerning diodes and LEDs.

3.2 Diodes and LEDs

Both diodes and LEDs operate in a similar manner with respect to the previously described bias effect (see Section 3.1.5.2). The major difference between the two is that an LED emits light in proportion to the forward current flowing through it. Thus, although they operate in a similar manner, the material and doping used for an LED results in the LED emitting light at a particular wavelength in proportion to the forward current flowing through the diode. Now that we have an appreciation of the basic differences between the diode and the LED, let's turn our focus to the operation of LEDs with respect to the bias effect.

3.2.1 LED Operation

Figure 3.6 illustrates the general operation of a generic diode/LED. The reason this author uses the term "generic" is that whereas all diodes and LEDs will not exhibit any operational capability in quadrants 2 through 4, in quadrant 1, the operational curve of forward current based on a forward bias voltage will vary depending on the type of the semiconductor material and its doping level. However,

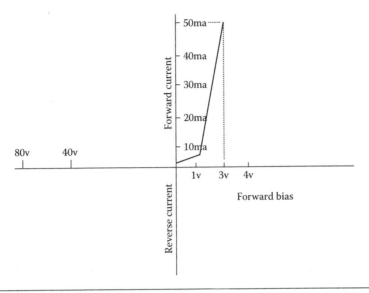

Figure 3.6 Diode/LED operation.

Figure 3.6 provides a good indication of how diodes and LEDs operate. That is, forward bias causes a forward current to flow through the device, resulting in an operational curve in the first quadrant. The operational curve varies according to vendor and vendor product and is usually included in the vendor's specification sheet, which indicates how the device operates.

3.2.2 Color of the Light Emitted by an LED

As previously discussed, the material used at the p-n junction as well as the degree of doping determine the wavelength of the emitted light. Unfortunately, there is no standard for LED color emission reference other than the emitted wavelength, which determines the color of the light. This means that because the human eye can distinguish a difference in certain colors of just a few nanometers, it's possible for LEDs emitting the same color to actually emit light at different wavelengths. For example, one LED manufacturer could emit yellow light at 590 nm, whereas a second LED manufacturer might use a wavelength of 580 nm. Because the human eye is capable of distinguishing a difference in yellow of approximately 1.5 nm, the two "yellow" LEDs can appear as two different colors. Similarly, this can occur for red and green primary colors and rationalizes why it is not a good idea to obtain LEDs from different vendors for a common application.

3.2.3 Light Production

In actuality, light represents a form of energy that is released by an atom. This form of energy has momentum but no mass, and the particles that provide the energy are referred to as *photons*, which represent the basic unit of light.

Photons are released due to the movement of electrons. As per our prior discussion of the atom (see Section 3.1.3), electrons move in orbits around the nucleus of an atom. As you might expect, electrons in different orbits have different amounts of energy, with the energy generally proportional to the distance of the orbit from the nucleus of the atom.

When an electron jumps from a lower orbit to a higher one, it requires a boost in its energy level. Similarly, when an electron drops from a higher orbit to a lower one, it will release energy. This energy is emitted as a photon, with a large energy drop resulting in the release of a higher-energy photon than a low energy drop. Because photons are emitted at defined frequencies, this also means that a large energy drop will be characterized by a higher-frequency photon than a low energy drop, which will be characterized by a lower-frequency photon.

As free electrons move across the p-n junction, they can fall into empty holes in the p-type layer. As this occurs, the electrons drop from the conduction band to a lower orbit around atoms, and release energy in the form of photons. Although the release of photons occurs in any diode, you can see the photons only when the diode consists of certain material that releases photons at predefined wavelengths. Otherwise, the photon's frequency will be invisible to the human eye as it will be in the infrared portion of the light spectrum.

To obtain visible light, the LED material needs to provide a wide gap between the conduction and the lower orbital bands, with the size of the gap determining the frequency of the photon that in effect defines the color of the light. As LED technology evolved, manufacturers began to use a variety of different materials in the semiconductor to generate peak spectral wavelengths that were associated with infrared and visible light. As we previously discussed the different types of materials used to develop LEDs as well as doping materials, we will conclude our discussion of light emission. However, readers are referred back to Chapter 1 in this book to obtain a detailed understanding of the use of different types of semiconductor material that can be used to emit light at different wavelengths which, in effect, produced LEDs that emit different colored lights.

3.3 Organic Light-Emitting Diodes

Often in the history of scientific developments, a person examining one technology will make a discovery that results in a new technology. Thus, the development of LEDs was a consequence of work with diodes. In addition, approximately 20 years ago, a Kodak scientist who was conducting research on solar cells discovered that transmitting an electrical signal through a carbon compound used in the cell resulted in the compound emitting light. This chance experiment resulted in the birth of organic light-emitting diode (OLED) technology. In the following sections, we will focus our attention on this approximately 20-year-old technology that has now moved from research into cell phones and TVs and has the potential to become a multibillion-dollar industry within a few years.

3.3.1 Overview

LEDs can be considered to represent small point light sources, which are based on the use of different semiconductor materials with different levels of doping. In comparison, OLEDs are based on organic (carbon-based) materials, which are manufactured in sheets and provide a diffuse area light source. Typically, a very thin film of organic material is imprinted onto glass. A semiconductor circuit is used to carry electrical charges to the imprinted pixels, which causes them to glow.

Today, OLED technology is used in a variety of display applications, ranging from cell phones and personal digital assistants (PDAs) displays to small TVs. Although OLED technology can be considered to represent an evolving technology on the basis of several market research reports published during 2008, it appears that the market for OLED displays alone will become a multibillion-dollar one within a few years.

3.3.2 Comparing Technologies

OLED is based on a process referred to as *electrophosphorescence*. That is, an OLED is an electronic device fabricated by placing a series of organic thin films between conductors. When an electrical current is applied, the organic material emits a bright light.

The organic material is referred to as a thin film because of the depth of the material. Typically, this material is less than 500 nm, which is 0.5 thousandths of a millimeter.

Compared to liquid-crystal displays (LCDs), OLEDs have a number of advantages. They are listed in Table 3.1. To understand some of them, let's briefly compare the two technologies.

3.3.2.1 LCDs versus OLEDs   LCDs are nonorganic, nonemissive light devices, which means that by themselves they do not produce any form of light. Instead, they will allow light to be reflected from an external source or provided by a backlighting system. Although the use of a backlight system provides an inexpensive method for displaying information, the backlighting accounts for approximately half the power required by an LCD system. Thus, OLED technology provides for a significant reduction in power requirements as a minimal amount of power is required to emit a bright light.

Other advantages of OLED displays over their backlighted cousin, LCD, are listed in Table 3.1. The increased brightness results from the fact that the OLED uses an electrophosphorescence process, whereas LCDs are primarily backlighted. Because an OLED uses a thin organic film instead of backlighting, the resulting display is lighter and brighter. In addition, OLED displays have a faster response time than LCDs, which enables the former to better display real-time information such as sports shows and complex gaming.

Although these advantages over LCDs are considerable, there is another advantage associated with OLEDs that has the potential to make them evolve into an eventual replacement for both plasma

Table 3.1 Display Advantages of Organic Light-Emitting Diodes (OLEDs) Over Liquid-Crystal Displays (LCDs)

Increased brightness	Wider viewing angle
Lighter weight	Enhanced durability
Faster response time	Broader operating temperature range

and LCD technologies. That advantage is the wider viewing angle associated with OLED displays. Along with a wider viewing angle, the ability to provide a distinct image even in bright light makes this display suitable for use in indoor and outdoor locations that periodically receive direct sunlight. Although some readers may not attribute much significance to these advantages, in all probability, such readers never realize that a large-screen TV in a big-box electronic storeroom produces screen clarity only because the retailer dimmed the lights where the TVs were displayed. To obtain the same clarity during the day at home, the homeowner has to close the drapes and curtains and dim the lighting, a situation that can be avoided once large-screen OLED TVs become available.

3.3.3 Types of Displays

Currently, there are at least six types of OLEDs, including active-matrix, passive-matrix, transparent, top-emitting, foldable, and white OLEDs. In the coming sections, we will focus our attention on two types of OLED displays, because they represent the primary areas in which both research and development of products are occurring. They are active-matrix and passive-matrix OLEDs.

The active-matrix OLED display is commonly and more formerly referred to as an *active-matrix organic light-emitting diode* (AMOLED), whereas the term *passive-matrix organic light-emitting diode* (PMOLED) is commonly used to reference a passive-matrix OLED display. Both displays have differences in their design and power consumption, which make them more suitable for certain applications than other types of displays.

3.3.3.1 PMOLED   A PMOLED is fabricated in a linear pattern, similar to a grid, with "columns" of organic and cathode materials superimposed on "rows" of anode material. In a passive-matrix environment, row and column electrodes are used to control the pixel at a given intersection. As pixels are turned ON and OFF in sequence, pictures form on the screen.

Passive-matrix OLED displays are well suited for text and icon displays used in all phones as well as automobiles and audio equipment. Although they are based on simple structures, they require relatively

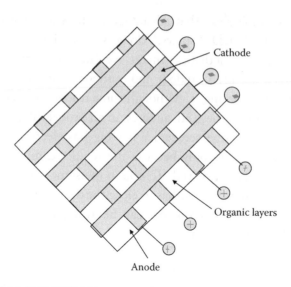

Figure 3.7 Passive-matrix organic light-emitting diode (PMOLED).

expensive drive electronics to operate effectively and also consume more power than that required by an active-matrix OLED. In fact, power analysis show that PMOLED displays are more practical for uses in which the display size is under 3 in diagonal or have less than 100 rows.

Figure 3.7 provides a general illustration of the construction of a passive-matrix OLED. Note that the organic layers are placed in a rib structure between the cathode and anode. Because both organic materials and cathode metal can be deposited into a rib structure, this facilitates the fabrication process and enables high-throughput manufacturing.

The operation of a PMOLED is relatively simple, requiring the application of a voltage to the row and column that corresponds to the pixel to be lit. Through the use of external controller circuitry, the necessary input power and video data signal is provided, with the video output displayed on the PMOLED panel via the successive scanning of all rows within a frame time, which is usually 1/60th of a second.

3.3.3.2 AMOLED   An AMOLED is similar to a PMOLED in that the cathode, organic, and anode layers are stacked on top of the substrate layer that contains the circuitry. However, unlike a PMOLED (where row and column electrodes are used to control the pixel at a

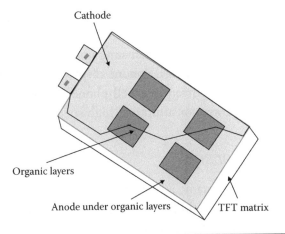

Cathode

Organic layers

Anode under organic layers

TFT matrix

Figure 3.8 Major components of an active-matrix organic light-emitting diode (AMOLED) display.

given intersection), every pixel is individually switched. To activate pixels directly, the circuitry delivers voltage to the cathode and anode materials, which stimulates a middle organic layer.

Another difference between active-matrix and passive-matrix OLEDs is the use of a thin film transistor (TFT) back plate on the active-matrix display. The TFT forms a matrix that is used to control both the brightness of the pixel as well as the pixel illuminated to form an image. In actuality, there are two TFT arrays per pixel. One array is used to start and stop the charge, whereas the second provides a constant electrical current to the pixel. Because a TFT array uses a very low amount of power and refreshes relatively fast, an AMOLED display consumes less power and is better for displaying quick action events than a PMOLED display. Thus, AMOLEDs are better suited for use in large-screen TVs, computer monitors, and billboards. Figure 3.8 provides a general illustration of the major components of an active-matrix OLED display.

3.3.3.2.1 AMOLED Applications The first commercial AMOLED was marketed in April 2003 by SK Display, a joint manufacturing venture established by Kodak and Sanyo. The first product produced by SK Display was used in the Kodak Easy Share LS633 digital camera display. This camera uses a 65,000 pixel color display, which was also used by Ovideon and NeoSol for use in personal media players during 2005. Due to financial reasons, SK Displays went out of business in late 2005. Currently, Sony and Samsung are the most aggressive AMOLED companies, with Sony planning several PDAs having AMOLED screens and Samsung investing $450 million in a factory that will produce mobile phone displays. The primary application for AMOLED currently is cell phone display. This is due to AMOLED's low power consumption, high speed, and thin profile.

3.3.4 Limitations of OLEDs

Although there are many advantages associated with the use of OLEDs, this author would be remiss if he did not describe and discuss some of their major limitations. The limitations include relatively short lifetime, the inability to manufacture OLEDs in large sizes, and ramp-up costs.

3.3.4.1 Lifetime of OLEDs Although OLED manufacturers have improved the materials and device structures over the past few years, the guaranteed hours of operation is currently an approximate maximum value of 15,000 hr prior to the brightness of the panel being reduced to half its initial value. Although this level of performance is acceptable for cell phones and similar consumer electronics with relatively small screens, it's currently inadequate for most TVs because they are typically sold with a medium 32 in. diagonal screen and are used extensively.

For example, consider the use of a cell phone versus a home TV. The cell phone may at most be used an hour a day, resulting in approximately 8120 hr of use per year. Because most persons replace their cell phone every 2 years or perhaps 3 or 4 years, if they do not use it frequently, a half-brightness life of 15,000 hr is typically not noticeable on a small cell phone display. Now let's compare the use of a large-screen home TV that might be installed on the wall of a den room. Typically, the TV might be on for 2–4 hr per day, resulting in a year of usage equivalent to between 16,000 and 32,000 hr of operation. Thus, the large-screen TV would lose half of its brightness in less than a year, a situation that would not be conducive to marketing large-screen OLED TVs. Thus, a significant improvement in the half-life of large OLED displays will have to occur prior to this type of display being used in large-screen TVs.

There are several areas manufacturers must confront in order to reach the 60,000 hr half-life of backlighted LCDs. First, new organic materials must be considered because current materials do not hold up under the driving current or the exposure to other materials in the device. Next, the cathode material is highly sensitive to air and even when sealed, this results in a degradation of OLED performance over time. But cell phone manufacturers can use OLED-based displays due to the fact that major users at 1 hr per day will take several years to reach the half-life of the display and will probably opt for a new phone prior to reaching the display half-life. In fact, most cell phone operators offer 2-year contracts, including a discount on a new cell phone, encouraging customers to both renew their contract and obtain a new phone. Thus, the effect of the half-life of OLED displays is minimal for the cell phone market. In comparison, the current brightness half-life of OLED displays and their inability to be fabricated in large sizes are major barriers to their use in TVs.

3.3.4.2 Fabrication and Ramp-Up Cost   As we will shortly note in this chapter (see Section 3.3.5), Sony introduced an OLED-based TV in late 2007. This first-to-market OLED-based TV has an 11 in. screen, which limits its market as consumers gravitate to larger TVs. Thus, the fabrication of larger size OLED displays is another current limitation of the technology. However, if you study the evolution of plasma and LCDs, you will note that over a period of approximately 8 years the average size of these displays almost tripled whereas their retail cost was reduced by a factor of 10. This required the investment of billions of dollars in factories constructed to fabricate the glass panels and electronics required to illuminate pixels. Thus, similar to plasma and LCD technology, the development of large OLED-based displays will require a considerable investment referred to as the *ramp-up cost*.

3.3.5 OLED TV

As discussed in Section 3.3.2.1, the key benefit associated with OLED displays over LCDs is that OLEDs do not require a backlight to function. This in turn significantly reduces the power requirement of the OLED display as well as makes it much thinner than an LCD.

Sony Corporation started selling the world's first OLED TV, XEL-1, in Japan in November 2007. The XEL-1 has an 11 in. display that is only 3 mm thick, which was placed on a pedestal with a flexible arm. The display provides a 960 × 540 pixel resolution, digital tuner, two 1 W speakers, HDMI, universal serial bus (USB), and Ethernet jacks in a package measuring 287 × 140 × 253 mm, with a weight of 2 kg or approximately 3.3 lb. According to news reports, Sony is initially limiting sales of the XEL-1 to Japan and its initial production run of 2000 units would be spread over 700 retail stores. In August 2008, the Sony store at The Forum Shops at Caesars was displaying the XEL-1 and it was now available for purchase in the United States.

The initial retail price of the Sony XEL-1 was set at 200,000 yen or approximately $1,700.00 at the conversion rate in late November 2007. If customer demand results in the initial production run being sold out, we can probably expect economics of scale to result in future price decreases. In addition, as Sony gains experience in the manufacture of larger size OLED displays, it's possible that this first-generation OLED-based TV will be supplemented with larger and longer-life OLED TVs.

In fact, shortly after the XEL-1 was announced, Sony displayed a 27 in. OLED prototype that had a 1,000,000 to 1 contrast ratio and was photographed from just about every possible angle by the electronic press due to its outstanding clarity. According to a Sony spokesperson, they are considering moving the 27 in. OLED TV into a production status.

Although there are several barriers including economics that preclude the widespread manufacturing of OLED TVs currently, it's important to note that the basic technology is only approximately 10 years old. In addition, during the past 10 years, a 42 in. plasma display, which cost approximately $15,000 and was subject to screen burn if left on for half a day or on a channel that had a logo displayed, can now be purchased for under a thousand dollars, includes electronics that minimizes the potential of screen burn, and provides a resolution double to triple that of earlier models. Thus, if OLED TV follows a similar evolution within the next decade, we can expect 50 in. OLED TVs that, due to their flexibility and thinness, could either be hung on a wall or simply pulled out of a retractable tube mounted horizontally on a wall and hung like a temporary painting. Although it's difficult to predict prices, especially many years from now, if OLED TV follows the plasma TV retail price curve, we can expect the 50 in. flexible screen to be under $1500 sometime in the next decade and perhaps even lower.

Although Sony was the first manufacturer to commercialize OLED TVs, other manufacturers in this evolving field have not been idle. During 2007, Seiko Epson Corporation, Nagano, Japan, developed the so far largest full-colored OLED TV using the manner in which conventional ink printing operates. The result was a 40 in. prototype display that was lightweight, thin, and had significantly high contrast and rapid response times.

When Samsung demonstrated an 82 in. LCD panel with a 180° viewing angle at the International Meeting on Information Display Conference held in Seoul, Korea, during 2005, the company was reported to be examining a full-color AMOLED display based on a white emitter with a red, green, and blue (RGB) color-filter array. This examination was to develop an alternative technology for large LCDs, which are relatively costly. Unfortunately, the use of RGB displays based on white emitters cause a portion of the white light to be absorbed by the color filters, reducing their advantage in power consumption in comparison to LCDs.

Along with the 82 in. LCD, Samsung displayed a very thin 40 in. AMOLED TV with a resolution of 1280 × 800 pixels, a contrast ratio of 5000:1, and a color saturation of 80 percent. Samsung also displayed a 17 in. AMOLED TV at the 2006 Korea Electronics Show, which was only 1.8 mm thick. It's possible that Samsung may eventually place this AMOLED TV, having a resolution of 1600 × 1200 pixels, a response time of 0.01 ms, and supporting a display of 262,144 colors, into production.

3.3.6 Other Markets

As previously noted, OLED displays are used in cell phones, automobile displays, and many types of consumer products. Concerning the latter, OLED displays are used in many types of MP3 players, digital cameras, and high-end audio equipment displays. According

Table 3.2 Comparing Large-Screen TV Technologies

PLASMA	LCD	OLED
Highest cost	Medium cost	Potentially lowest cost
Requires large power	Consumes less power	Consumes least amount of power
Uses backlight	Uses backlight	Self-emissive
Displays deeper black	Displays fewest colors	Displays more colors than LCDs
Screen burn potential	No screen burn potential	No screen burn potential
Half-life of 60,000 hr	Bulb replacement at	Currently red and green elements only last
	Half-life of 60,000 hr	Half-life between 10,000 and 40,000 hr
		Blue half-life at or under 10,000 hr

to one forecast, the market value for OLED panels is expected to rise to $3.5 billion by 2012, without considering the potential growth in OLED-based panels for large TVs. If OLED large-screen TVs follow the price declines of plasma and LCD TVs, it's quite possible that once OLED average lifetime increases to 60,000 hr, the economics associated with the technology could rapidly replace other types of large TVs. If this occurs, the value of OLED panels could easily exceed $10 billion by 2012.

To obtain an appreciation of OLED TV functions in comparison to plasma and LCD-based TVs, Table 3.2 provides a general comparison of the technologies. In examining the three columns in this table, you will note the absence of tube-based TVs. Although this author believes that such TVs will still be manufactured by 2012, they will be a minor niche market and will not compete with large-screen TVs that will be primarily wall mounted.

3.4 LED Drivers

Until now we have simply noted that we can use one or more resistors to control the current flowing through a circuit containing one or more LEDs. In the following sections, we will enhance our knowledge of the control of current in an LED circuitry by turning our attention to LED drivers. After briefly discussing the rationale for LED drivers, we will define a generic driver. Using this information as a base, we will then examine some of the different types of drivers and the applications they can support.

3.4.1 Rationale for Use

Previously in this book, we noted that LEDs are current-driven devices. As the current increases, the brightness of the LED increases. Thus, its brightness is proportional to the forward current.

There are essentially two methods that can be used to control the forward current flowing through an LED. The first method involves the use of the LED's voltage–current curve to determine the level of voltage required to generate the desired forward current. In fact, in Chapter 1 of this book, we examined how this was accomplished by using a resistor and Ohms law to determine the current that would flow through an LED.

That is,

$$I = \frac{V_s - V_f}{R}$$

where

V_s = voltage at source
V_f = forward voltage
R = resistance

Although it's common to use a resistor in simple LED applications, for more sophisticated applications there are several drawbacks to this method. First, any change in the LED forward voltage will create a change in the LED current. Thus, a slight voltage change that is within a specified voltage tolerance due to temperature or manufacturing changes can result in a forward current change. For example, consider the circuit shown in Figure 3.9 in which a 6 V battery and 120 Ω resistor are used to drive an LED that has a forward voltage of

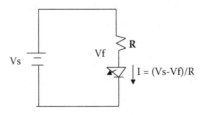

Figure 3.9 The use of a resistor to limit forward current.

3.6 V, resulting in a forward current of 20 mA. Suppose the voltage changes to 3.8 V. Then, the forward current will become

$$I = \frac{6 - 3.8}{120} - 18.33 \text{ mA}$$

Thus, an approximate 5.7 percent ((3.8 − 3.6)/3.8) change in the forward voltage results in a 8.5 percent ((20 − 18.33)/20) change in the forward current. Thus, a change in the forward voltage can result in a larger percentage change in the forward current. This can result in an LED burn or a significant reduction in the half-life of the device, depending on the change in the forward voltage.

A second method used to regulate LED current is obtained by driving the device with a constant current source. Through the use of control circuitry, a constant current source is provided or "driven" to the LED. This constant current eliminates variations in current due to changes in the forward current, which results in a uniform or constant level of LED brightness. Thus, driving an LED with a constant current is a better method of regulating LED current.

3.4.2 Using PWM

If we take the use of driver circuitry a bit further, we can obtain an understanding of how LEDs can blink and change their level of brightness. To do so, the driver circuit can provide a pulsing current, in effect use pulse width modulation (PWM), with different duty cycles of the current waveform to transition the LED between its ON and OFF states; the time spent between and in each of the two states provides the impression that the LED is being dimmed, in a flashing mode, or being both dimmed and in a flashing mode of operation.

The rationale for using PWM instead of regulating the amount of current is how many LEDs react to a low current. Although the light output of an LED is proportional to drive current over most of the LED's current range, as the current level significantly decreases, the device can either flicker or change color. Thus, although both methods (PWM and regulating the drive current) can be used to adjust LED brightness, PWM provides the ability to pulse an LED as well as is the preferred method for dimming LEDs. Now that we have an appreciation for the use of PWM in drivers, let's define the LED driver.

3.4.3 Driver Definition

The definition of an LED driver is a bit tricky as there are many types of drivers that are used to perform different functions. In general, we can define an LED driver as a stand-alone control circuitry or a self-contained power supply with control circuitry that provides an output matching the electrical characteristics of the LEDs to be controlled. If the LED is to be dimmed or pulsed ON and OFF, the driver would then include PWM circuitry.

3.4.4 Driver Connection

Unfortunately, there is no standard that defines LED driver connections, with different manufacturers sometimes varying the characteristics and connections used for their drivers. In general, an LED driver commonly includes at least four connectors. Two connectors are for power input, whereas the other two are for the LED. If the driver supports PWM, it will normally include two additional connections. These connections are typically used for a dimming pot or control connection.

3.4.5 Types of Drivers

Currently, semiconductor companies market a broad line of LED drivers, which are designed to support a variety of applications. Such applications range from automotive (where LEDs are used in interior and exterior lighting) to display (where LEDs provide a backlighting capability). In addition, different types of drivers are used on the basis of the topology of the application, where LEDs can be mounted in series or parallel. In spite of the preceding we can place LED drivers into five general categories: boost or step-up LED drivers, buck or step-down LED drivers, buck-boost LED drivers, multitopology drivers, and pump LED drivers. In the following sections, we will briefly note the characteristics of each type of driver.

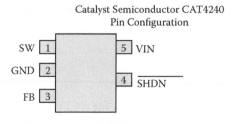

Figure 3.10   Pin configuration of the CAT4240 LED driver.

3.4.5.1 Boost LED Drivers   A boost or step-up LED driver is designed to boost voltage. In doing so, some semiconductor products will have a feedback mechanism that uses the forward voltage (V_f) of the LED as a mechanism to adjust the driver's boost voltage to a fixed value above the highest measured V_f. By adjusting the boost voltage, the power dissipation between the boost voltage and the forward voltage is minimized, which also minimizes the power dissipated in the driver. Regardless of the type of boost LED driver, they all function as a dc-to-dc step-up converter, which provides a constant current for driving one ore more LEDs. If the driver does not include PWM circuitry, LED dimming can be performed by controlling the dc voltage or via a logic signal.

3.4.5.1.1 Types of Boost Drivers   The most common type of boost driver uses an inductor as it is very efficient over a wide range of output voltages. For illustrative purposes, consider Figure 3.10, which shows the pin configuration and use of the Catalyst Semiconductor's CAT4240 boost LED driver. This LED driver can drive six LEDs configured in series. Readers who require additional information concerning this driver should visit the vendor's Web site at www.catsemi. com. In addition, it's a good idea to check a vendor's Web site as product specifications can change. Later in Section 3.4.5.5, we will discuss a second type of boost LED driver referred to as a *pump driver*.

Concerning the pins of the Catalyst driver shown in the left portion of Figure 3.10, their general functions are listed in Table 3.3.

3.4.5.2 Step-Down LED Drivers   A step-down or buck LED driver is a dc-to-dc converter that is used to lower the provided input voltage to a voltage that supplies an acceptable constant current to the

Table 3.3   Pin Functions of the CAT4240 LED Driver

PIN #	NAME	FUNCTION
1	SW	Switch pin. This is the drain of the internal power switch.
2	GND	Ground pin. Connect to the ground plane.
3	FB	Feedback pin. Connect to the last LED cathode.
4	SHDN	Shutdown**pin (logic low). Set to high to enable driver.
5	VIN	Power supply input

LED. Depending on the type of step-down driver, PWM circuitry may be included to support dimming as well as flashing. Currently, a step-down LED driver is commonly used for automotive interior and exterior lighting, as well as ambient and architectural lighting. These application areas have a voltage source far above that required to operate one or more LEDs, thus the step-down or buck LED driver support LEDs very well in such environments.

3.4.5.3 Buck-Boost LED Drivers   Now that we have an appreciation for the general operation of buck and boost LED drivers, the name buck-boost prefixing another type of LED driver explains its capability. That is, a buck-boost driver provides the ability to either step-up or step-down the source voltage. As you might expect, the cost of this capability exceeds drivers designed only to perform one function. However, the dual capability provides more flexibility for designers when prototyping an application.

3.4.5.4 Multitopology Driver   Most LED drivers are designed for supporting one or more LEDs configured in series. However, there are situations in which LEDs need to be configured in parallel. To support both topologies, you might expect vendors to offer multitopology drivers. However, this term is actually misleading as these drivers are limited to supporting one topology. For example, one vendor currently offers an inductorless LED driver for LEDs in parallel and an inductor-based LED driver for LEDs in series under the term "multitopology LED driver."

3.4.5.5 Pump LED Driver   The last category of drivers we will cover in this section can be considered as a variation of the boost LED driver. That driver is the pump or charge pump LED driver.

Table 3.4   Comparing Current-Limiting Methods

	ADVANTAGES	DISADVANTAGES
Resistor	Inexpensive; one large component required	Cannot accurately control circuit Current will vary with supply voltage. Power disruption in the resistor requires it to be accurately sized
Driver	Allows dimming through current control Permits dimming via PWM Permits compensation for the effect of temperature changes Can accurately regulate LED current	More expensive than a resistor May require a heat sink Requires careful design to avoid EMI

Boost drivers can typically be subdivided into two based on the manner in which voltage is adjusted. The first type of boost LED driver is based on an inductor, whereas the second type is based on the capacitor-based charge pump converter. The charge pump LED driver gets its name from the use of a charge pump converter to boost voltage. The charge pump converter is efficient when the output voltage is a multiple of the input voltage; however, this places some restrictions on how the voltage can be applied. In comparison, the boost dc-to-dc converter, which was previously discussed in Section 3.4.5.1, used an inductor for conversion and not only has no output voltage restrictions but, in addition, is more efficient than a charge pump converter over a wide range of output voltages.

3.5 Summary

In this chapter, we expanded our knowledge concerning how current can be limited in an LED by examining LED drivers. Table 3.4 provides a general comparison of the use of resistors and LED drivers, noting the advantages and disadvantages associated with each device.

4

LEDs and Lighting

At first, the title of this chapter may appear a bit odd for as you already know, light-emitting diodes (LEDs) emit light. However, instead of completely focusing on the manner in which LEDs emit light, this chapter will primarily focus on their use in lighting applications. Such applications range the gamut from the use of LEDs in flashlights and backlighting liquid-crystal displays (LCDs) to the introduction of LED-based lightbulbs that can be expected to eventually replace fluorescents in a manner similar to how fluorescent lightbulbs are now replacing filament-based lightbulbs.

In this chapter, we will primarily focus our attention on the use of LEDs to provide an alternative to conventional filament-based lighting as well as the more modern fluorescent lighting. In doing so, we will first discuss the rationale for the replacement of filament-based lightbulbs that, as strange as it might seem, is not based on economics, which consumers tend to ignore, but primarily on congressional mandate and utility subsidization of the cost of compact fluorescent lightbulbs. Once we obtain an appreciation for the utility industry subsidization of compact fluorescent lightbulbs and the congressional mandate for replacing filament-based lightbulbs, we will briefly examine the economics associated with their replacement. In doing so, we will not only examine the cost of operating filament-based, fluorescent, and LED bulbs, but in addition look at their life cycle cost.

In the second section of this chapter, we will turn our attention to the LED-based lightbulb. We will describe and discuss a relatively new type of LED referred to as a *high-brightness* (HB)

LED. In doing so, we will note how HB LEDs and their verbal cousins referred to as *ultrabright* (UB) LEDs are adding new applications that can be satisfied through the use of LEDs.

4.1 Rationale

Common sense tells us that the replacement of an "energy hog" by a similar device that has a life an order of magnitude greater and operates using a quarter of the electricity of the device to be replaced should attract consumers when the price differential is not exorbitant. This economic situation has existed for approximately half a decade between incandescent or filament-based lightbulbs and fluorescent lightbulbs, yet until recently the incandescent lightbulb retained a vast majority of the market for lightbulbs. Perhaps, a part of the reason for the lack of success of fluorescents to capture any appreciable market share of the lightbulb replacement market was the fact that, until recently, consumers viewed lightbulbs as low-cost disposable items. After all, you could go to a grocery store, drug store, or convenience store and pick up a four-pack of 40, 60, 75, or 100 W lightbulbs for a bit under $2. However, if you waited for one of the frequent two-for-one sales, you might have stocked up on enough lightbulbs to last for a long time.

4.1.1 Incandescent Lightbulbs

Common incandescent lightbulbs have not changed their design nor energy efficiency in over 100 years. An incandescent lightbulb uses an electrical current that is passed through a thick filament located within a closed glass bulb. The filament is heated to the point where it emits light and the glass bulb is used to prevent oxygen from reaching the filament, which could then be destroyed by oxidation. Unfortunately, only about 5 percent of the electricity used is turned into light, whereas the remainder is turned into heat. If you live in an area having many hot and humid summer months, the heat from the incandescent bulb adds to the load on the air conditioner, bumping up the cost of energy to light and cool one's home or business.

Although regular incandescent lightbulbs have an average life of 750–1000 hr, most are turned on only for a fraction of a day. For a typical working family with children attending school, lightbulbs may be illuminated at worst, continuously between 4 PM and 11 PM, or 7 hr per day. Assuming a minimum life of 750 hr, a person purchasing a four-pack has enough bulbs to last well over a year for one fixture. Thus, a second reason there is no rush to replace incandescent lightbulbs is the "inventory" of bulbs most homeowners keep in drawers and closets.

A third reason for the reluctance of homeowners to replace incandescent lightbulbs is economics. Until recently, fluorescent lights were sold for $3 or more per bulb. Although the $3 bulb lasts longer and reduces one's electric bill, the one-time cost for happy homeowners going to a store for a four-pack might convince them to opt for conventional incandescent lightbulbs, especially when they are more concerned about paying to fill their gas tanks.

4.1.1.1 Economics of Use   To illustrate the cost associated with using incandescent filament-based lightbulbs, let's assume we are located within an area in the United States where the cost of electricity is 10 cents (¢) per kilowatt-hour (kWh). If we assume the bulb has a 750 hr life and a four-pack can be bought for $2.00, then the one-time cost per bulb is 50¢. If we assume we are using a 100 W bulb, then over its average 750 hr life it can be expected to use 750 hr × 100 W/hr (75000/1000) or 75 kWh. At a cost of 10¢/kWh, the operating cost of the bulb becomes $7.50. When we add the purchase cost of the bulb, the total cost becomes $8.00 to use a 100 W bulb over its 750 hr life, assuming that the cost per kWh is a dime. We can perform similar computations for 75, 60, and 40 W lightbulbs that also have a 750 hr life expectancy. The results of these computations for four popular incandescent lightbulbs are listed in Table 4.1. From examining the

Table 4.1   Incandescent Lightbulb Lifetime Costs

WATTAGE	ELECTRICITY	ONE-TIME COST	TOTAL COST
100	7.50	0.50	8.00
75	5.63	0.50	6.13
60	4.50	0.50	5.00
40	3.00	0.50	3.50

entries in the table, as you might expect, the higher the wattage of a lightbulb, the higher the consumption of electricity, which in turn increases the total lifetime cost of the incandescent lightbulb.

Although the cost of electricity used in computing the operating costs shown in Table 4.1 was set at 10¢/kWh, readers should note that the cost of electricity can considerably vary according to location. Some areas in the United States that use inexpensive hydropower may have a cost per kWh significantly under 10¢/kWh, whereas other areas that use gas turbines for extra generation during peak hours may have a blended cost considerably above 10¢/kWh.

4.1.2 Compact Fluorescent Lightbulbs

Over the past few years, three significant events occurred, which can be considered to represent driving forces facilitating the penetration of compact fluorescent light (CFL) bulbs into the market for lightbulbs. These forces include a reduction in the cost of fluorescent bulbs, utilities subsidizing the cost of the bulbs, and the Federal 2007 energy bill that includes rules for lightbulbs which traditional incandescent bulbs cannot meet.

4.1.2.1 Cost Reduction CFL bulbs date to the early 1990s, when they were manufactured either as a circular or U-shaped device that required a special fixture. By the mid-1990s, manufacturers started making spiral bulbs that were actually invented at General Electric in the 1970s. The use of spiral bulbs resulted in the emission of more light than linear bulbs. In 2001, Home Depot and other big-box hardware stores began to carry compact fluorescent bulbs. However, their initial cost of roughly $11 each was approximately 20 times the cost of conventional incandescent lightbulbs.

Due to product design, some of the earlier problems associated with CFLs, such as reaching full brightness over a period of time and not using significantly less energy over an extended life than projected, were resolved while manufacturing economics of scale resulted in lower prices. Today, most stores that carry conventional incandescent lightbulbs also stock CFL bulbs and the price of the latter can be obtained in a three-pack for as low as $4.95 to $7.95 at many outlets. Thus, due to a ramp-up in production as well as improvement in product design, the cost of CFL has fallen from a ratio of approximately 20:1 to 4:1 in comparison to the cost of an incandescent lightbulb.

4.1.2.2 Utility Subsidization According to the California Energy Commission, approximately 37 percent of electricity use per California household during 2006 was attributed to lighting. Recognizing that it could be less expensive to reduce demand than to construct and operate new electric generation plants, several utilities including Pacific Gas and Electric Company (PG & E), San Francisco, California, are subsidizing CFL bulbs. This subsidization was not altruistic, but due to a mandate.

California's three major investor-owned utilities were mandated during the summer of 2007 to reduce their combined energy use by the equivalent of three electric generation plants. On the carrot side of the mandate, each utility can earn up to $180 million over a 3-year period, whereas if they fail to meet the goals of the state's conservation plan, they would face significant financial penalties.

Because lighting consumes more electricity in California homes than cooling, refrigeration, and ventilation combined, California utilities decided that it was possible to reduce the consumption of electricity attributable to lighting to obtain over half of their state-mandated conservation gains. To accomplish this, PG & E and other utilities have spent millions on subsidizing CFL bulbs. According to an article published in the January 9, 2008, issue of the *Wall Street Journal*, bulbs that cost between $5 and $10 each during 1999 and that today can retail for several dollars apiece are available in California for between 25¢ and 50¢. On the downside, some of the subsidized bulbs are showing up on the shelves of out-of-state stores and in eBay auctions. However, this situation is probably not unexpected as it's next to impossible to police the sale of lightbulbs. The important thing to consider is that utilities are transforming the market for compact fluorescent lighting. Even if they should stop subsidizing the cost of CFL bulbs tomorrow, which is highly unlikely, economics of scale would result in a significant cost reduction that, although above the 25¢ to 50¢ subsidized price, would be a significantly lower cost than if the subsidization program never occurred. In fact, the January 2008 issue of *Living in South Carolina* included an article that indicated that the state's electric cooperatives backed the National Action Plan for

Energy Efficiency. This plan is intended to save $100 billion on nation-wide energy bills as well as reduce carbon emissions. One aspect of the series of methods by South Carolina's electrical cooperatives to hold down electric bills is to place energy efficient CFL bulbs in the home of every electrical cooperative member in the state. Over a 10-year period, the savings could equal the energy used by 35,000 homes if just seven million bulbs are installed. According to the article, South Carolina cooperatives plan to invest up to $10 million per year toward renewable energy and energy efficiency measures.

4.1.2.3 The Federal 2007 Energy Bill   A third factor that has an impact on the replacement of incandescent lightbulbs is the energy bill passed by Congress during 2007, which President George W. Bush signed into law. Under this energy bill, incandescent lightbulbs would be gradually phased out of use, beginning in 2012, with 100 W bulbs removed from sale. A year later, 75 W bulbs will be removed from sale, followed by the more popular 60 and 40 W bulbs during 2014.

By sending incandescent lightbulbs to join the dildo bird, consumers will be forced to purchase other types of lightbulbs. Although CFL bulbs are currently being subsidized by certain electric utilities, this is not to say that eventually LED-based lightbulbs will not wind up being subsidized. However, such bulbs will not only have to be price competitive with CFL bulbs but in addition provide significantly more savings to the consumer. This may be difficult to achieve as certain manufacturers of halogen and other types of lighting have announced they are working on methods to increase bulb efficiency. For example, Phillips plans to market a halogen light during the fall of 2008 that will be significantly more efficient and three times longer lasting than incandescent bulbs; however, they will be initially more expensive than CFL and incandescent lightbulbs. As manufacturing ramps up and economies of scale result, the energy efficient halogen light could become a competitor to LED-based lightbulbs, which would then have two competitors: CFL bulbs and halogen lightbulbs.

4.1.2.4 Economics of Use   If we compare a CFL bulb to an incandescent lightbulb, we will note that the efficiency of the CFL makes its cost significantly less compared to the life of the bulb. For example,

Table 4.2 Comparing Incandescent and CFL Lightbulbs

CHARACTERISTIC	INCANDESCENT	CFL
1. Bulb life (hour)	750	7500
2. Number of bulbs required	10	1
3. Cost of bulb(s)	$2.50	$4.00
4. Watts consumed	100	20
5. Electrical operating cost	$75.00	$15.00
6. Total cost	$77.50	$19.00

a CFL typically has a life expectancy 10 times that of a conventional incandescent or about 7500 hr. During that time, it consumes approximately one-fifth of the energy of an incandescent bulb. Let's compare the economics of use associated with a 100 W incandescent lightbulb and a CFL light designed to produce an equivalent amount of light but which consumes 20 W of power. This comparison is shown in Table 4.2.

In examining the entries in Table 4.2, note that the electrical operating cost (5) is computed by multiplying the bulb life (1) by the number of bulbs required (2) by the watts consumed (4) and then dividing the result by 1000 to obtain kWh. That result was then multiplied by 10¢/kWh. Finally, the total cost (6) was obtained by adding the cost of the bulbs (3) and the electrical operating cost (5). As indicated in Table 4.2, the total cost of a CFL lightbulb is approximately one quarter that of an incandescent lightbulb.

4.1.2.5 Disposal Problems Unlike incandescent and LED lightbulbs, the disposal of CFL can result in the release of toxins. This results from the fact that each compact fluorescent bulb has approximately 5 mg of mercury sealed inside the glass tubing. If a CFL bulb is dropped and shatters, it should be cleaned up right away and the room should be aired out for 15 min as a precaution. However, when tens of thousands of CFL bulbs make their way into landfills, their breakage could result in mercury contaminating soil and underground water unless the landfill has precautions in place. Unfortunately, as of 2008 it appears no government agency has considered that within a few years tens of millions of CFL lightbulbs will reach the end of their lives. Prior to this occurring, recycling programs will be needed to route disposed CFL bulbs so that their mercury is removed prior to the bulb landing in a landfill. Just how this will work is anyone's guess at the present time. In fact, although some states currently prohibit the disposal of CFL bulbs in one's garbage, if placed in a recycle bin chances are very high the bulb is not recycled and reaches a landfill where, when crushed, it will release its toxins.

4.1.3 LED Lightbulbs

In the following sections, we will turn our attention to the LED-based lightbulb. In doing so, we will first discuss some of the key characteristics of this lightbulb, after which we will examine its operating cost in comparison to the use of incandescent lightbulbs and CFL bulbs.

4.1.3.1 Purchase Considerations Although LED lightbulbs can now be located on the shelves of big-box retailers as well as on numerous Internet shopping sites, their purchase requires careful consideration of numerous factors. These factors include their lumen output, light direction, initialization or turn-on time, ability to be used with a dimmer, operating life, watts consumed, heat dissipation, and quality of light.

4.1.3.1.1 Lumen Output Until recently, the lumen output of an LED lightbulb was a fraction of that of incandescent bulbs. For example, a typical 75 W incandescent bulb has an output of 1050 lumens (lm). In comparison, most LED lightbulbs were limited to between 20 and 80 lm. Thus, it required at least 14 LED lightbulbs to provide the same light output as provided by a single incandescent lightbulb. Recently, manufacturers have attacked this problem by adding more LEDs per bulb. Instead of 12, 24, or 36 LEDs per bulb, with approximately 2–3 lm of light per LED, new lightbulbs are being manufactured with 150 or more LEDs, increasing their lumen output to approximately half of a conventional 100 W incandescent lightbulb.

Because LED lighting is highly directional, LED lightbulbs can be used in certain locations to provide as much directional lighting as a conventional incandescent. Lightbulbs with a small number of LEDs are currently better suited for low-level light intensity applications, such as night lights, garden path lighting, and similar applications.

4.1.3.1.2 Direction of Emitted Light   An LED is limited to projecting light in one direction. Thus, generation of multidirectional light requires numerous LEDs pointed in different directions to be fabricated within a lightbulb. As you might expect, there is a cost relationship between the number of LEDs used and the cost of the lightbulb, with a 150-LED bulb costing considerably more than a bulb with 36 LEDs.

Currently, the light output of LEDs can be characterized as cold and not very powerful. This is changing due to the introduction of high-brightness (HB) LEDs that will be discussed in Section 4.2.

4.1.3.1.3 Initialization or Turn-On Time   The initialization or turn-on time represents the time taken from turning on power to a light until it reaches a steady state of illumination. Whereas an incandescent lightbulb has a near-instantaneous turn-on time, until recently CFL lightbulbs could take several seconds to reach a steady state level of illumination, especially in cold temperatures. However, in the past 2 years, the slow flickering and the yellow-white color have been replaced by near-instant-on CFL. In comparison, LED bulbs lighting is near instantaneous, similar to incandescent lightbulbs.

4.1.3.1.4 Dimmer Use   Currently, only incandescent and some types of CFL lightbulbs are suitable for use in circuits with dimmer controls. Unfortunately, light dimmers do not work with LED bulbs unless there is at least one incandescent bulb in the circuit. Even then, as the dimmer control is activated, the change in the resistance that dims lighting can shorten or burn out the LED bulb. In this author's opinion, time is required for LED bulb manufacturers to develop the technology to enable LED bulbs to both incorporate more LEDs as well as work correctly on dimmer circuitry.

4.1.3.1.5 Operating Life   Most LED lightbulbs can operate for approximately 60,000 hr or over 7 years of continuous operation 24/7, which explains why they are used in traffic lights where the use of red, amber, and green can allow a life of over 14 years, as one color is shown only for a portion of the time. In comparison, conventional incandescent lightbulbs have an operating life of 750 hr, whereas

CFL lightbulbs may exceed 7500 hr of life. It's important to note that the operating life of bulbs can vary based on their design and use. Concerning the latter, constantly turning on and off a CFL or LED light can drastically shorten its life.

4.1.3.1.6 Watts Consumed   Currently, there are two primary reasons for considering the use of LEDs. One is its long life, which can average between 60,000 and 100,000 hr, depending on the type of the LED. A second key reason for using LEDs is their low power consumption.

When discussing the use of LEDs in a lighting environment, it's important to note that an LED bulb consumes approximately one-tenth of a watt per LED in the bulb. Because LED bulbs come in many sizes and shapes, there is no standard bulb and hence no standard LED bulb wattage. However, at an approximate power consumption of one-tenth of a watt per LED, this means that a 150-LED bulb will consume 15 W of electricity, which is 20 percent of the power consumption of a conventional 75 W incandescent lightbulb. Thus, the use of LED lightbulbs can considerably reduce your electricity bill and save the cost of the numerous replacement bulbs you avoid having to purchase.

4.1.3.1.7 Heat Dissipation   Unlike incandescent and CFL bulbs, the diodes in an LED lightbulb convert just about all the power consumed into visible light. Thus, there is almost no heat generated by an LED lightbulb. This can be especially important in areas of high heat and humidity, where heat from lighting places an extra load on air conditioners.

4.1.3.2 Quality of Light   The LED emits light perpendicular from the diode. Thus, light flows straight out from the bulb but can be spread by the use of a lens. By adding LEDs and fabricating them so that they point in different directions downward and by the use of a dispersing lens, it is possible to spread light within a circular area. However, most LED lightbulbs are currently used as small directional lights or night lights. An exception to this usage is the use of high-power or high-brightness (HB) LEDs (which we will discuss in the next section), which can be used as recessed lights, for track lighting, and similar applications.

When attempting to ascertain the economics associated with the use of LED lightbulbs, there are two key issues that need to be addressed. The first issue involves the cost and type of an LED lightbulb. During 2008, when this book was researched, this author noted over 100 types of LED lightbulbs in the market. Although some LED lightbulbs appeared to be similar to one another, upon closer inspection there were variances in the number of LEDs used in a bulb, lumen output, power consumption, and other metrics. Because we are focusing on the comparison between incandescent and CFL lightbulbs, this author only considers LED bulbs that would fit in a standard electrical socket. A second key issue is the operating life of an LED bulb. Because the life of an LED bulb could be approximately 100 times that of an incandescent lightbulb, an operating cost comparison over the life of each type of bulb would not be meaningful unless the number of equivalent incandescent and CFL bulbs was considered. Thus, a carefully thought-out cost comparison approach to obtain an economics of use was required.

The approach selected to examine the economics associated with the use of LED lightbulbs was obtained by considering the number of incandescent and CFL bulbs that would be used to obtain the minimum 60,000 hr life span of an LED lightbulb. Assuming that a standard incandescent lightbulb has a 750 hr life, 80 such bulbs would provide a life expectancy equivalent to one LED lightbulb. Similarly, using a life expectancy of 7500 hr for a CFL lightbulb would result in the need for 8 such bulbs to reach the life expectancy of an LED lightbulb.

A new problem arose when attempting to match an LED lightbulb to incandescent and CFL lightbulbs. This problem was concerned with the number of lumens each bulb outputs. Because LED bulbs are more focused, their output in lumens are less than that of other types of bulbs; but the lighting they produce in certain areas, such as on a kitchen table, may be similar to the output of other bulbs. For comparison purposes, the author selected a 9 W LED bulb designed to replace a 70 W incandescent lightbulb. Although the retail price of $80 was considered expensive, this LED lightbulb consisted of 150 individual LEDs and provided 594 lm of output, nearly equivalent to that of a regular 75 W incandescent lightbulb and its CFL lightbulb replacement when considering the focus of the directional light of the LED bulb.

Table 4.3 Operational Life Cost Comparison

CHARACTERISTIC	INCANDESCENT	CFL	LED
1. Life (hour)	750	7500	60,000
2. Number of bulbs	80	8	1
3. Cost of bulbs	$20.00	$24.00	$80.00
4. Watts consumed	100	20	9
5. Electrical operating cost	$800.00	$120.00	$54.00
6. Total cost	$820.00	$144.00	$134.00

Table 4.3 indicates the operational costs associated with operating three nearly equivalent types of lightbulbs: incandescent, CFLs, and LEDs. Note that due to differences in lumens, life expectancy, and wattage, they are nearly equivalent and not equal. However, the information in the table permits a reasonable comparison between the uses of each type of light source, once again assuming that the cost of electricity is 10¢/kWh.

In examining the entries in Table 4.3, note that the electrical operating cost (5) is computed by multiplying the bulb life (1) by the number of bulbs required (2) and the watts consumed (4); and then dividing the result by 1000 to obtain kWh. This result is then multiplied by 10¢/kWh. Finally, the total cost (6) is obtained by adding the cost of the bulbs (3) and the electrical operating cost (5). As indicated in Table 4.3, the total cost of a CFL lightbulb is approximately the same as that of an LED lightbulb at the present time. Because the LED bulb consumes approximately half the power of the CFL bulb, when manufacturing efficiencies enable a drop in the price of LED bulbs, it will become economically viable for uses at home.

4.2 High-Brightness (HB) LEDs

Less than 7 years old, HB LEDs have the potential to become a "killer application," which enables LED lightbulbs to be used for general-purpose illumination both inside and outside the home and office. In the following sections, we will turn our attention to this relatively new type of LED, noting its evolution and how recent developments, including the ability of HB LED bulbs to be directly connected to alternating current (ac) instead of using an ac–dc (direct current) converter, result in a considerable increase in efficiency while producing two to three times as much light as a regular conventional powered LED.

4.2.1 Overview

HB LEDs actually date back to 2001 when LEDs began to be created based on semiconductor epitaxial growth techniques. One of the semiconductor growth techniques that emerged as a leader approximately 7 years after the initial development of HB LEDs in the laboratory is a method referred to as *metal-organic chemical-vapor deposition* (MOCVD). Under this technique, the base wafer, which is silicon or germanium, is heated on a graphite susceptor to a very high temperature that enables the wafer to accept doping. Initially, propylene is passed over the wafer to clean the carbon resulting from heating the wafer. Next, a semiconductor material is generated by doping the silicon wafer. Doping with aluminum and boron results in p-type and n-type semiconductors, respectively; other elements can also be used.

4.2.1.1 Metal-Organic Chemical-Vapor Deposition System

Figure 4.1 illustrates the basic structure of a MOCVD system. Note that a variety of different gases are used, such as N_2 to fill the reactor, H_2 as a carrier of the metal organic source, SiH_4 to dope a wafer as an n-type semiconductor; gases are used based on the type of HB LED to be manufactured.

The growth of a p- or an n-type semiconductor is a multistep process. First, the substrate is cut into slices and then cleaned. This is followed by the creation of a vacuum in the reactor and the flow of

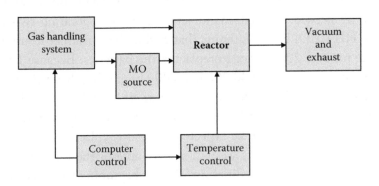

Figure 4.1 Basic structure of a metal-organic chemical-vapor deposition (MOCVD) system.

H_2 gas as a carrier of the metal-organic (MO) source. Finally, under computer control a program will control such parameters as the flow rate, timing, chamber pressure, and valves in the system.

4.2.1.2 Initially Developed HB LEDs   Perhaps the first HB LEDs you are familiar with are those that have been used in traffic signals and warning lights. You can note them as an array of small yellow or red light points in comparison to the solid light produced by a single conventional bulb covered and brightened by a reflecting mechanism. The yellow and red spectrum HB LEDs were among the first to reach the market, and today, they are considered primitive with respect to advances that have occurred in the development of HB LEDs.

However, even though considered primitive by today's HB LEDs, the use of LEDs in traffic lights has been steadily increasing over the past decade. Due to their extremely long life and lower power consumption in comparison to conventional lighting, the use of LEDs in traffic signals save cities and municipalities at least tens of millions of dollars each year.

4.2.1.3 Utilization   As HB LEDs evolved, they found their way into various types of lighting applications, ranging from residential uses to alternatives for signs and displays with the latter including electronic billboards. If you travel on a major highway, you will probably notice the use of LED-based electronic billboards. Similarly, HB LEDs are commonly used in most sporting events, even though the type of sport may differ. Nowadays, an increasing number of flashlights no longer have a conventional miniature bulb that usually burns out just when it's needed. Instead, manufacturers are incorporating two and three HB LEDs into flashlights that enable the consumer to use the device for over 20 hr prior to replacing the batteries. Because LEDs have a life expectancy of 40,000 to 60,000 hr, it is highly unlikely that the consumer will have to worry about anything other than the life of the batteries in the flashlight. As we move toward the next decade, LED-based lighting can be expected to rapidly replace many types of conventional lighting. As we will shortly note in the following sections, cities and municipalities are already beginning to replace conventional outdoor lighting with HB LEDs. As manufacturing ramps up, the unit cost of LED-based lighting can reasonably be expected to decline. This in turn will result in an increased market penetration of LED-based lighting.

4.2.2 Fabrication Forms

There are two forms of HB LEDs that have evolved from the use of a series of semiconductor materials. The first form is based on the use of indium, gallium, and phosphide or AllnGaP, which is pronounced as "alan-gap." The use of this group of semiconductors results in the creation of orange-red, orange, yellow, and green HB LEDs. A second solid-state HB LED form uses indium and a gallium nitride compound and is referred to as *InGaN*. Pronounced as "n-gan," this solid-state compound is used to create blue, blue green, true green, and (when combined with yellow phosphor) white light.

Currently, Nichia Corporation, Anan, Japan, is the world's largest manufacturer of GaN-based LEDs. In 1993, Nichia developed the world's first blue HB LED. Since then, Nichia has expanded its offerings to a full range of LEDs in varying sizes and levels of brightness. Nichia HB LEDs have an estimated life between 60,000 and 70,000 hr and are able to withstand large variances in temperature and moisture, making them useful for numerous types of outdoor applications such as in billboards and signs on tall buildings.

4.2.3 ac versus dc Power

LEDs are normally considered to be dc devices that operate from a low-voltage dc power source. For low-power applications that use a small number of LEDs, such as in mobile phones, the power is supplied by a dc battery. However, for other applications such as a linear strip lighting system that is installed around the outside of a building or as a string showing the aisle exit walkway within an airplane, the use of dc may not be suitable. This is because dc suffers from increasing losses as distance increases. To compensate for such losses, the use of higher drive voltages as well as additional regulators that waste power is required.

Because ac performs better over distances and is well established and simple to use, we can consider the holy grail of LED-based lighting to be the operation of LEDs from a main supply, such as

120 V ac. This requires electronics between the supply and the LED bulbs, which provides a dc voltage capable of driving the LEDs. This resulted in the development of the ac–dc converter and drivers that can be considered the "fixture," no pun intended, of first generation HB-LED lightbulbs.

Although such products represented several orders of magnitude in their level of brightness over conventional LEDs, the use of ac–dc drivers typically results in a loss of approximately 15 percent. Clearly, a new approach that would enable LEDs to operate directly from an ac power supply would be beneficial. In addition to avoiding the loss associated with ac–dc drivers, the electronics required to convert ac to dc can also be eliminated. Thus, you obtain in effect a double advantage if you can operate LEDs from an ac power supply. However, you should not confuse the ability to power LED lightbulbs from ac with the majority of LED lightbulbs in the market during 2008. Such bulbs that screw into an ac socket include miniaturized circuitry that converts ac to dc prior to delivering power to the LEDs in the bulb.

4.2.3.1 Seoul Semiconductors In 2007, Seoul Semiconductor, Seoul, Korea, introduced the world's first semiconductor lighting source that uses ac current directly, without conversion. Marketed under the name Acriche, a compound word with "Acro" meaning top and "Riche" from French meaning richness, there were 10 HB LEDs in the product line in early 2008. Five operated at 2 W, while the other five products operated at 4 W.

Each of the Acriche products can be considered as the electrical equivalent of two long strings of HB LEDs placed in series. This results in the product becoming nonpolar, with light emission occurring on both halves of an ac cycle. Because the strings that are fabricated are formed from many p-n junctions in series, the total forward voltage of each string is relatively high, approaching the input operating voltage. Concerning the input operating voltage, the Acriche product line supports operating voltages of 100V, 110V, 220V, and 230V. Perhaps one trade-off to the Acriche product line is the life expectancy of the LEDs, which was listed as "more than 35,000" hours on the Seoul Semiconductor Web site located at www.Zled.com/en/product/prd/acriche.asp

In comparison to the "more than 35,000" product life specified for Acriche LEDs, most manufacturers specify a life expectancy of at least 60,000 hr for HB LEDs that require ac–dc drivers. Thus, it appears there is currently a trade-off between driver efficiency gained by the ac solution and the life expectancy of an LED.

4.2.3.2 Lynk Labs   Shortly after the introduction of the Seoul Semiconductor Acriche series of ac-powered LEDs, Lynk Labs, Inc., Elgin, Illinois, introduced its XyLite™ ac LED light engine technology. XyLite supports both low- and high-voltage ac, and through the use of modules, provides *X*- and *Y*-dimensional building blocks that enable OEM designers to combine as few as two LEDs in a single assembly or package with matching driver technology for the specific ac-powered LED.

XyLite's basic module is 14 × 15 mm and consumes 1.5 W. At 14 × 45 mm, the module consumes 4.5 W. Using Lynk Labs' constant voltage ac drivers, LEDs are driven in opposing parallel circuit configurations at different frequencies depending on the application. For example, a high-frequency and low-voltage driver would be used to drive an ac LED device or assembly that matches the constant-voltage driver, whereas other devices and assemblies could be designed to connect directly to main power or for such low-voltage applications as landscape lighting via the use of a transformer.

The Lynk Labs' ac approach employs an even number of LEDs in a circuit that contains a capacitor as shown in Figure 4.2. Similar to the Seoul Semiconductor approach to direct ac, Lynk Labs' system is designed so that both half cycles of the ac wave are used. The purpose of the capacitor is to drop the voltage and deliver the required current to the LEDs based on the voltage and frequency of the ac source input to the capacitor. In addition, the capacitor insures that a

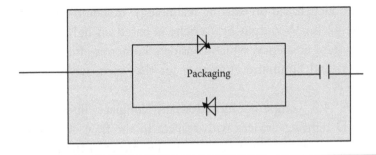

Figure 4.2   Lynk Labs' ac-LED module.

constant current is delivered to the LEDs as well as isolates an LED from other LEDs in the system and from the driver if a failure should occur.

Currently, the U.S. Department of Energy is extending its Energy Star program to include HB-LED-based illumination applications. This extension will enable both homeowners and businesses to receive tax credits and other incentives for acquiring a variety of HB-LED types of lighting. In Section 4.2.5, we will discuss this program with respect to the evolving HB LED.

4.2.4 HB-LED Output

To understand why HB LEDs may eventually replace a variety of incandescent and CFL lighting, we need to focus our attention on its light output in luminous flux or luminous intensity. Recall that the unit of luminous flux is lumen, abbreviated as lm. The lumen is defined as 1/683 W of optical power at a wavelength of 555 nm. Luminous intensity takes into account the direction of light emitted by an LED and is related to luminous flux. That is, one lumen per steradian equals one candela (cd) of luminous intensity.

For comparison purposes, consider a 60 W incandescent light-bulb. It uses 60 W of energy, and according to its package, emits approximately 1000 lm. Thus, its luminous efficiency is 1000/60 or approximately 16.6 lm/W. In comparison, Seoul Semiconductor's white-warm Acriche ac-powered HB LED features 42 lm/W, whereas other LEDs yield over 50 lm/W.

According to the U.S. Department of Energy's Lighting Technology Roadmap, it's expected that HB LEDs should be able to produce 150 lm/W by 2012 or approximately triple that of currently available LED products based on existing technology. It should be mentioned that the 42 lm/W output by Acriche is based on light output. The actual system efficiency, which considers luminous efficiency, ballast efficiency, and luminaire efficiency, for the warm white Acriche is 39.9 lm/W.

Figure 4.3 compares incandescent, compact fluorescent, and HB-LED lighting devices with respect to the flow of photons. As you can see from the illustration, only the output of the HB LED is directional, enabling the output to be focused.

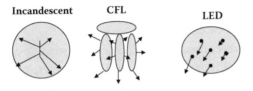

Figure 4.3   Comparison of lighting devices.

4.2.5 Energy Star Program Developments

Earlier in this chapter (Section 4.2.3.2), we briefly mentioned the Energy Star program and LEDs. In this section, we will expand our knowledge of the program.

On September 12, 2007, the Department of Energy finalized its criteria called "Program Requirements for Solid State Lighting Luminaires." The Energy Star program for solid-state lighting establishes a two-category approach. Category A addresses near-term applications, in which SSL technology can be appropriately applied. In effect, Category A covers devices that are currently available and states that a Category-A-compliant recessed lighting fixture or "downlight" must have a luminous efficiency of 35 lm/W. Category B establishes targets for a wide range of emerging applications that are expected to reach the commercial market within the next 3 years. Thus, Category B will come into effect once solid-state lighting technology becomes more mature. At that time, Category A will be dropped and Category B will become the sole basis for the Energy Star criteria with respect to solid-state lighting.

In effect, Category B represents future performance targets. One interesting target concerns luminaire efficacy that is listed as being greater than or equal to 70 lm/W, which is double the value of 35 lm/W specified for Category A. Once Category B comes into effect, LED lighting will rival or exceed the most efficient lighting currently available. For example, the commonly available T8 fluorescent and electronic ballast produces approximately 100 lm/W and represents one of the most efficient types of lighting currently in the market. High-quality fixtures for the lamp-ballast system are about 70 percent efficient; this results in a 70 lm/W luminaire efficacy. Although at the beginning of 2008 no solid-state lighting product could match this level of luminaire efficacy, by the time Category B standard is met

LED lighting should match or exceed the luminaire efficacy of high-performance fluorescents.

As energy costs continue to rise, we can reasonably expect that lighting efficiency will increase in importance. In fact, a study performed in California (which was previously mentioned in Section 4.1.2.2) as well as the U.S. Department of Energy's estimate attribute approximately 22 percent of all energy consumption as being used for indoor and outdoor lighting. If within a few years LED efficiency increases as expected, energy consumption will be significantly reduced. As a general approximation, doubling the efficiency of LEDs and their use in place of several hundreds of millions of light sockets now using incandescent and CFL lighting could reduce total energy consumption by 10 percent. Although a 10 percent reduction may not appear to be significant, this is equivalent to approximately two million barrels of oil per day. Or to put this reduction into perhaps a more meaningful set of terms, as one barrel is the equivalent of 48 gal, this implies that the United States could reduce energy consumption by 96 million gallons of gas equivalent per day. In addition to reducing the use of electricity, the low power consumption of LEDs would also reduce carbon emissions and eliminate the necessity to construct many generating plants. According to the Department of Energy, the rapid adoption of LED lighting could eliminate approximately 260 million metric tons of carbon emissions and avoid the necessity to construct 133 power plants over the next 20 years. Thus, there are significant advantages associated with the development, production, and installation of HBNN-LED lighting products.

4.2.6 Outdoor Lighting Developments

In September 2007, Cree, Inc., Durham, North Carolina, one of the major innovators who manufacture semiconductors that enhance the value of LED lighting, achieved the highest efficiency of 129 lm/W for a cool-white LED and 98 lm/W for a warm-white LED. Although this occurred in the laboratory, it verifies the expectations of the U.S. Department of Energy that over the next several years significant advances will occur in the efficiency of LEDs. In fact, in November 2007 Cree's XR-E series of LEDs were significantly improved. The advanced version of the XR-E series now produces white light output of up to 250 lm, which represents a 56 percent improvement over earlier technology. The XR-E has a cool-white output with up to 114 lm at 350 mA and an efficacy of over 90 lm/W. In addition, Cree developed a "Warm" series of LEDs for indoor uses. The Warm series provides up to 80.6 lm at 350 mA with an efficacy of up to 67 lm/W. Cree's XR-E series of high-power LEDs were voted as the most significant "leapfrog" technology story of 2007 by readers of Electronic Design magazine. Perhaps, the ultimate proof of the utility of the XR-E series is that the company replaced the lighting in its lobby and parking lot with its white LEDs.

4.2.7 Cities Discovering LEDs

One of the costs associated with a city's budget is the power consumed by street lighting. When you consider the cost of energy, the total operational cost including the maintenance cost associated with replacing burnt-out lightbulbs indicates that LED lighting can pay for itself between 6 and 8 years. In fact, the city of Raleigh, North Carolina, estimated that converting lighting to LEDs would provide an investment payback in 7 years. According to the City of Raleigh, an initial investment of $42,000 would reduce electrical consumption by $2803 per year. In addition, bulb replacement and other maintenance costs would be reduced by $3325 per year due to the longer life of LEDs. Together, this would save the city $6128 per year, resulting in a (3525/42000) × 100 or 14.5 percent return on the city's investment not counting the time value of money, which is insignificant due to several years of relatively low interest rates. In an era where CD rates hover around 4–5 percent for a 10-year commitment, the ability to achieve a 14.5 percent return is truly phenomenal. What makes the return even more outstanding is the fact that you can be sure the price of lightbulbs and electricity will continue to increase over the next 7 years. This means that savings from reduced electrical demand and less maintenance will continue to increase. Thus, the actual return on the city's investment should increase, whereas its payback period can be expected to decrease.

In addition to the City of Raleigh, other cities and municipalities are installing LEDs to replace traditional street lighting. For example, the city of Ann Arbor, Michigan, plans to install over 1000 LED streetlights before 2010. According to city officials, they are projecting a 50 percent reduction in the cost of energy and a 3.8-year payback on their initial investment in LED-based lighting.

In addition to reducing the costs of electricity and maintenance, the replacement of traditional street lighting by LEDs has an environmental effect. This is because the reduction in electrical consumption results in a reduction of the generation of greenhouse gases by coal, oil, and gas generation plants. This is turn reduces the emission of carbon dioxide as well as other pollutants into the atmosphere.

4.2.8 Lighting Science Group

In concluding this chapter, this author would be remiss if he did not mention the role of the Lighting Science Group Corporation, one of the leading suppliers of LED lights and lighting fixtures in the United States. OSRAM Opto Semiconductors, Regensburg, Germany, is a leading manufacturer of LEDs that are used in many of the Lighting Science Group's products. Located in Dallas, Texas, the Lighting Science Group Corporation has multiple patents pending in the areas of bulb design, power management, and the manufacturing process associated with LEDs. In fact, Lighting Science Group was selected by the city of Raleigh to provide light fixtures for the city's parking garages as Raleigh uses LED technology to replace high-intensity discharge lamps in street and facility lighting.

The Lighting Science Group currently markets a series of LED lamps, controllers and dimmers, software, and peripheral products. The company has also developed customized lighting solutions that resulted in LEDs being used on the DMX bridge in Korea, the MTQ bridge in Quebec, Montreal, numerous office buildings and inside hotel lobbies, and in a performing arts center. Currently, Lighting Science Group markets a series of six LED lamps whose functionalities are summarized in Table 4.4.

In examining the entries in Table 4.4, it's interesting to note that the R series LED lamps provide a range of savings from 23 percent ($((50 - 15)/15) \times 100$) for the R38 lamp to 73 percent ($((50 - 6)/50) \times 100$) for the R20 lamp. These computations take the difference between the equivalent incandescent bulbs wattage and the LED's wattage, which

Table 4.4 Functionalities of Lighting Science Group LED Lamps

| MODEL | PRIMARY UTILIZATION | LUMENS | | LIFE (HOURS) | EQUIVALENCY | ENERGY CONSUMPTION (WATTS) |
		WARM WHITE	COOL WHITE			
G25	Hard-to-reach locations	100	120	50,000	15 W incandescent	4.5
MR16	Accent lighting	300	35	35,000	20 W halogen	5.0
R16	Flood, track, and spot lighting	300	35	35,000	40–45 W incandescent	5.0
R20	Commercial accent lighting	325	40	50,000	Incandescent	6.0
R30	Commercial accent lighting	550	75	50,000	50–65 W incandescent	13.0
R38	Displays of perishable goods	650	87	50,000	50–65 W incandescent	15.0

is then divided by the LED bulb wattage, and the result is multiplied by 100.

The use of the R series lamps for the variety of applications listed in Table 4.4 can considerably reduce energy consumption at home and office. This is especially true for commercial applications in which lights may be kept on for extended periods of time, such as in department and big-box stores as well as 24/7 supermarkets. Now that we have an appreciation for some of the products of the Lighting Science Group, we will conclude this chapter with a brief discussion of the manufacturer of LEDs used in many of the Lighting Science Group products. That manufacturer is OSRAM Opto Semiconductors GmbH.

4.2.9 OSRAM Opto Semiconductors

OSRAM Opto Semiconductors of Regensburg, Germany, was founded as a joint venture between OSRAM GmbH and Infineon Technologies during 1999. The founding of the company resulted from Siemens divestment of its semiconductor operations, enabling OSRAM GmbH, one of the world's leading manufacturers of lighting products, to eventually acquire OSRAM Opto Semiconductors as a wholly owned subsidiary. The name OSRAM is from the words Osmium and Wolfram (German for Tungsten), because both these elements were commonly used for lighting when the company was founded. Currently, OSRAM Opto Semiconductors is the world's second largest manufacturer of optoelectronic semiconductors that are used for illumination purposes. Today, OSRAM Opto Semiconductors manufactures a variety of HB LEDs in various white tones as well as in numerous colors that are used in such lighting sectors as room lighting, street lighting, architecture lighting, department store and hotel lighting, to name but a few.

One of the key products of OSRAM Opto Semiconductors used by the Lighting Science Group is its high-performance LED marketed under the registered name of Golden DRAGON. Golden DRAGON LEDs are available on tape and reel, with the LED being manufactured with a flat top to enable "pick-and-place" machinery installation. The LED includes an integrated heat spreader that according to OSRAM Opto Semiconductors provides a far superior level of thermal performance than obtainable with standard LEDs.

The Golden DRAGON LED can operate at current levels in hundreds of milliamperes. Versions of Golden DRAGON LED are available based on the uses of InGaAIP and InGaN. Recall that InGaN technology is used to generate colors from blue at a wavelength of 460 nm to true green at 528 nm as well as phosphor-based colors such as white. In comparison, InGaAIP is used to generate colors from green at 570 nm to super red at 632 nm. InGaAIP has a forward voltage between 1.8 and 2.3 V, depending on the color. In comparison, InGaN has a much higher forward voltage, which is between approximately 3.2 and 3.8 V, again with the forward voltage depending on the color.

Golden DRAGON LEDs based on InGaAIP can be obtained in amber red and yellow colors. Products based on InGaN can be obtained in blue, verde green, true green, and white colors. InGaAIP LEDs operate from 100 to 750 mA, whereas InGaN LEDs can operate up to 500 mA. Because the high current results in a large amount of power that needs to be dissipated, a heat spreader is incorporated to enhance power dissipation. This in turn enables the LED to generate more light because a cool LED emits more light than a hot component, which could be one of the reasons for its selection in some of the Lighting Science Group's products.

5

LEDs in Communications

In this chapter we will focus on the manner in which LEDs are being used to transmit information. Although most readers will be aware of the role of lasers in transmitting information via fiber-optic cable, to paraphrase the late Rodney Dangerfield, "LEDs get no respect" yet play an important role in the wonderful world of modern communications.

From infrared (IR) remote control to high-speed Ethernet networking, LEDs provide a low-cost, viable transmission capability that is the subject of this chapter. As we proceed through this chapter, we will focus attention on the manner in which LEDs are being used in a variety of communication applications. Because readers have different backgrounds, prior to explaining the use of LEDs in a particular communications application, the underlying technology or set of technologies associated with a particular communications method will be explained. Hopefully, by the end of this chapter, readers will have a new appreciation for LEDs and, unlike the late Rodney Dangerfield, you will also have a new respect for the use of LEDs in different types of applications that require a communications capability.

5.1 Remote Control and Infrared LEDs

Perhaps one of the most important inventions of the 20th century for operator convenience was the remote control. First developed as a mechanism to change television channels on black and white TVs, today there are literally hundreds of different types of remote controls. For example, in addition to controlling televisions, VCRs, DVD players, and audio components, one can also purchase fans, garage doors, space heaters, and even notebook computers that work with an infrared remote control. In this section, we will first discuss what the term *infrared* means. Using this information as a base, we will then examine IR technology and the role of LEDs in this technology.

5.1.1 Overview

The term *infrared* refers to the portion of the electromagnetic spectrum that is not visible to the human eye. As a brief refresher, the portion of the electromagnetic spectrum that is visible ranges from 400 nm (violet) to 700 nm (deep red). Wavelengths shorter than 400 nm or larger than 700 nm are not normally visible to the human eye.

5.1.2 The Infrared Region

As a matter of definition, wavelengths shorter than 400 nm are referred to as *ultraviolet*, whereas wavelengths greater than 700 nm are referred to as *infrared*. Infrared is further subdivided based on its wavelength. Although the IR portion of the electromagnetic spectrum is from 700 nm to approximately 1300 nm, the spectrum from 700 nm to 950 nm is referred to as *near-infrared*. It is this area of the IR spectrum in which modern IR LED emitters and detectors operate.

5.1.2.1 Rationale for Use The use of the IR region instead of visible light was determined to represent a more effective method for controlling remotely performed operations. That is, detectors would otherwise be sensitive to visible light variances. Thus, the sun peaking out of a cloud and shining into a room or the turning on of a halogen lamp could result in the false trigger of a visible light sensor.

5.1.2.2 Frequency and Wavelength In Chapter 2, we noted that frequency is the reciprocal of wavelength. That is,

$$F = c/\lambda,$$

where c = 299, 792, or 458 nm/s, which is the speed of light in a vacuum.

Table 5.1 Regions of the Electromagnetic Spectrum

REGION	WAVELENGTH (Å)	WAVELENGTH (CMS)	FREQUENCY (HZ)
Radio	$>10^9$	>10	$<3 \times 10^9$
Microwave	10^9–10^6	10–0.01	3×10^9–3×10^{12}
Infrared	10^6–7000	0.01–7×10^{-5}	3×10^{12}–4.3×10^{14}
Visible	7000–4000	7×10^5–4×10^{-5}	4.3×10^{14}–7.5×10^{14}
Ultraviolet	4000–10	4×10^{-5}–10^{-7}	7.5×10^{14}–$3 \times 10^{17**}$
X-rays	10–0.1	10^{-7}–10^{-9}	3×10^{17}–3×10^{19}
Gamma rays	<0.1	$<10^{-9}$	$>3 \times 10^{19}$

The frequency for a given type of electromagnetic radiation is used to determine its position on the electromagnetic spectrum. Thus, it is common to find charts of the electromagnetic spectrum expressed in terms of wavelength or frequency, or more commonly, in terms of both metrics.

Table 5.1 provides a summary of the frequency and wavelength associated with the major regions of electromagnetic radiation in the known frequency spectrum. As you will note from the entries in Table 5.1, high-frequency electromagnetic waves have a relatively short wavelength, whereas low-frequency waves have a relatively long wavelength.

Recall from physics that the angstrom was named in honor of the Swedish physicist Anders Jonas Angstrom, who created a spectrum chart of solar radiation that expressed the wavelength of electromagnetic radiation in multiples of one ten-millionth of a millimeter. This unit of length, which is 1×10^{-10} m or one-tenth of a nanometer, became known as the *angstrom*, abbreviated as Å.

In addition to listing the regions of the electromagnetic spectrum in tabular form, we can also show the regions in a plot based on frequency, wavelength, or both metrics. Although the latter requires a bit of artistic skill, this author decided to show the regions in terms of wavelength.

Figure 5.1 illustrates the electromagnetic spectrum, showing the major regions from gamma rays that have relatively short wavelengths to radio waves that have relatively long wavelengths. Note from Figure 5.1 that visible light is a very small portion of the electromagnetic spectrum. Also note that the IR portion of the spectrum has a greater wavelength, and thus, a lower frequency than visible light. Although we cannot hear IR, if the signal is strong enough, we can feel IR as heat.

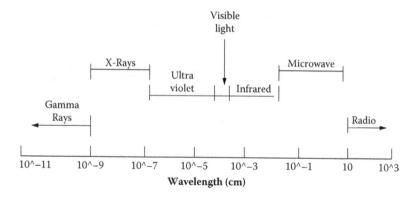

Figure 5.1   The electromagnetic spectrum.

5.1.3 Evolution in the Use of IR

IR came be considered a viable mechanism for use in remote control during the early part of the 1990s, when the Infrared Data Association (IrDA) was formed. IrDA is a nonprofit organization established to promote interoperability among IR devices as well as set global standards for transmitting information via infrared. Currently, it has over 150 corporate members and provides a mechanism for the certification of products under the monikers "IrReady" and "IrSimple Shot."

Currently, IrDA is developing specifications for the transfer of information via an IR port at a data rate of 1Gbps. Once products are available, the contents of a CD can be transferred from one device to another via IR technology in under a second. Because modern remote controls are used to send data, let's first turn our attention to the manner in which an IR remote operates.

5.1.4 IR Remote Operation: IR Port

In this section we will discuss the use of the IR port built into many electronic devices, including all devices that need the capability to be controlled remotely. Once this is accomplished, we will briefly review the different types of IR devices.

An IR device such as a remote control or IR receptacle on a computer, digital visual radio (DVR) player, or another device contains a small, dark window referred to as the *IR port*. It is used to transmit and receive IR beams; however, prior to transmitting actual data, the device with the IR port converts the binary data to be transmitted into sequences of IR light.

5.1.5 Types of IR Devices

There are a range of IR devices manufactured today. Five of the major types of IR devices include emitters, detectors, photo interrupters, photo reflectors, and transceivers.

5.1.5.1 Emitters
An emitter type of IR device has a single function: to transmit IR. The emitter transmits a narrow spectrum of IR radiation. One common example of an IR emitter is the common TV remote control. In communications, we would refer to an emitter as a simplex or one-way transmission device.

5.1.5.2 Detector
Similar to an emitter, an IR detector has a single function. However, unlike the emitter, which transmits IR radiation, the detector is designed to receive IR radiation. An IR detector can be either a photodiode or a phototransistor. Often, they are used with optical IR filters to minimize their response to visible light or to radiation outside the desired range of IR radiation the device is designed to receive.

5.1.5.3 Photo Interrupter
A photo interrupter represents a transmission-type sensor that combines an emitter and detector into a single package. When fabricated, the emitter and detector are placed on the same axis across from one another, separated by a small gap. A device such as a printer or plotter that uses a photo interrupter is designed so that the object to be sensed interrupts or breaks the "beam" between the emitter and detector, enabling the system to respond. The photo interrupter is commonly used in printers, fax machines, plotters, and copying machines to detect the presence or absence of paper and its position in certain products.

In addition to the preceding, compact photo interrupters are also found in cameras, where they are used to detect precise focus and zoom lens positions. Because the photo interrupter is designed to detect breaks in the IR LED beam, they are also commonly used in different types of industrial machinery, such as an industrial press and industrial stamping machine, which must manufacture parts with a high degree of precision.

5.1.5.4 Photo Reflector   A photo reflector is similar to a photo interrupter in that both consist of an emitter and a detector in a single package. However, in a photo reflector, the emitter and detector are placed next to each other on the same plane and not across from each other, which represents how a photo interrupter is fabricated.

In a photo-reflector-based system, the object to be sensed reflects the IR radiation back from the emitter to the detector, resulting in the system being able to respond to the IR beam. Both the physical placement of the emitter and detector in a common package as well as the angle of the beam generated by the emitter are optimized for sensing at a predefined distance. Although you might not think of IR when you visit a public restroom, the paper towel dispensers and automatic water faucets represent two common applications based on the use of photo reflector IR devices.

5.1.5.5 IR Transceiver   An IR transceiver, which represents an acronym for "transmitter/receiver," combines an emitter and sensor either with one or two lenses within a single housing. The use of a single lens permits a relatively small form factor to be achieved, whereas the emitter is usually positioned vertically above the sensor.

The miniature IR transceiver is commonly used on an Ethernet network interface card for transmission over multimode fiber. However, LED transceivers are also available for a wide range of applications. For example, one can use an IR transceiver with a universal serial bus (USB) interface to transmit and receive IR codes. Some products include software drivers for different laptop and desktop computers that enable many types of remote controls to be used to control the computer. When combined with a high-speed Internet connection, a USB cable, and a computer connected to a television in effect, one can control the download and display of movies and video clips, such as YouTube "movies." Now that we have a general appreciation for the different types of IR devices, let's turn our attention to the ubiquitous TV remote control.

5.1.6 TV Remote Control

A TV remote control includes an IR LED that is used to transmit in one direction, providing in effect a simplex or one-way transmission capability. Most TV remote controls use a directed IR beam produced by LEDs that have a moderate cone angle. This explains why you need to point your remote in the general vicinity of the TV. In addition, the IR beam has a range of approximately 30 ft, which explains why persons in a modern contemporary home that has several open rooms without walls may not be able to control a TV located in one room from an area in another room.

5.1.7 The IR Signal

The IR signal generated by a TV remote control is typically modulated using amplitude shift keying (ASK). ASK is used to modulate a carrier signal.

5.1.7.1 ASK Modulation

ASK represents a form of modulation in which digital data in the form of binary ones and zeros are transmitted as variations in the amplitude of a carrier. Perhaps, the easiest and most popular form of ASK modulation is to have either amplitude or no amplitude, depending on the binary value of the digital signal.

Figure 5.2 illustrates an example of ASK, in which the ASK signal in the lower portion of the referenced figure is based on the binary sequence of digital data to be modulated, which is shown in the upper portion of the figure.

One of the key advantages of ASK modulation, no pun intended, is that it's relatively easy to implement. To do so, you only need a

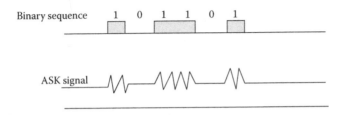

Figure 5.2   Amplitude-shift keying modulation.

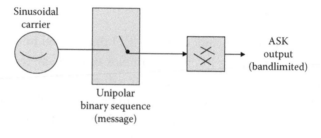

Figure 5.3 Generating an ASK signal.

sinusoidal carrier, unipolar binary data source, and switch. Figure 5.3 illustrates the generation of an ASK signal.

One of the problems associated with ASK modulation occurs owing to the sharp discontinuities in its waveform. As previously shown in Figure 5.2, ASK can go from zero to some height and then back to zero based on bit duration. These discontinuities would result in a wide bandwidth that could significantly increase the error rate at the receiver. Thus, a mechanism to reduce the bandwidth is required. That mechanism can be implemented by bandlimiting (pulse shaping) the message prior to modulation or by limiting the ASK signal after modulation. In Figure 5.3, the ASK output is shown as bandlimited after modulation. Concerning the carrier that is modulated, most remote control units use a 36 to 40 kHz carrier and transmit data at a rate between 100 to 2000 baud or bps.

5.1.7.2 FSK Modulation   In place of ASK, some remote control devices use frequency-shift keying (FSK). It was one of the earliest modulation methods developed for communications as it's very simple to implement. Under FSK, a frequency shift occurs based on the binary value of digital data to be transmitted. That is, f_1 (the first frequency) is used to represent a binary 1, whereas a second frequency f_2 is used to represent a binary zero.

5.1.8 Interference

There are numerous sources of interference that can cause problems when using an IR remote control. Some problems can result from other IR sources such as some audio systems that continuously broadcast an

IR signal. Other causes of interference can result from the ballast of fluorescent lights and machinery that generate short bursts near or within the 36–40 kHz frequency commonly used by IR remote controls. To limit such interference, many manufacturers turned to the use of higher IR carrier frequencies.

Although some IR systems use carrier frequencies that fall into the megahertz (MHz) region, most IR remote control devices typically use a 36–40 kHz modulated square wave for communication. The square wave is sent to the IR transmitter, which uses an IR LED. To obtain an appreciation of how a TV remote control operates, let's turn our attention to a generic device.

5.1.9 Inside a TV Remote Control

Figure 5.4 illustrates the front view of a basic generic TV remote control. For simplicity, this author did away with the numerous buttons on modern TV remote controls and instead used a control that simply changes channels and volume, has a single toggle ON/OFF power button, and buttons for muting and displaying changes on a TV screen. This facilitates both the drawing and review of the operation of the remote without having to spend a lot of time drawing many buttons that are basically irrelevant to the operation of an LED inside the remote control.

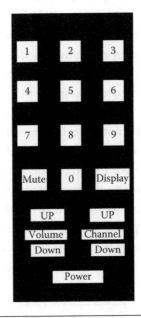

Figure 5.4   A basic generic TV remote control.

5.1.9.1 Operation
The function of the remote control is to translate key presses into IR light signals that are received and acted on by a television. If you take apart the TV remote control, which is not recommended by this author, you will see a printed circuit board (PCB) that contains several components mounted on the board as well as

battery contacts and what appears to be a series of wires that represent copper etching on the circuit board.

5.1.9.2 Printed Circuit Board The printed circuit board (PCB) is a thin piece of fiber glass, which does not conduct electricity. On the circuit board are thin copper wires that are etched onto its surface and enable data to flow between the components mounted on the board.

Figure 5.5 illustrates a generic TV remote control PCB viewed from the bottom after the removal of the back of the remote control housing. In actuality, most remotes only have a battery cover to access the battery compartment. Thus, you may need to unscrew or pry loose the connectors to open the remote, and neither action is recommended by the author. Instead, we will simply use the view provided in Figure 5.5 to note the relationship of the major components on the PCB to the LED.

When viewed from the bottom, you cannot see the vast majority of the copper etching as it resides on the opposite side of the circuit board. This placement results from the fact that circuitry is required to convey the pressing of keys on the remote control. Returning our attention to what is essentially the rear of the PCB in the lower right, you will note the battery compartment that provides power to the components on the PCB. At the upper right is an integrated circuit that is used to detect when a key is pressed and then transmits a coded sequence to the transistor, which modulates and amplifies the signal prior to it being sent to the LED. Each button on the remote has a contact point so that when the button is pressed, it pushes the contact point onto the circuit board. When this occurs, data flows to an integrated circuit that defines the button that was pressed.

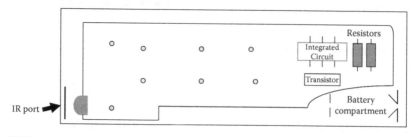

Figure 5.5 A TV remote control printed circuit board.

5.1.10 Remote Control LEDs

Earlier in this chapter, we noted that wavelengths greater than 700 nm are referred to as being in the ultraviolet region of the electromagnetic spectrum. Specifically, the region from 700 nm to 950 nm, which is referred to as *near-infrared*, is the area in which most modern IR LEDs operate.

Because IR LEDs do not produce visible light, their output is not measured in millicandelas, which is the term used to measure the light output of visible lighting. Instead, their output is measured in milliwatts. Another difference between LEDs that produce visible light and IR LEDs is their current and forward voltage ratings. IR LEDs have a lower forward voltage and a higher rated current than visible LEDs. This difference is due to the properties of the material used at the p-n junction of the LED diode.

5.1.10.1 Wavelengths and Fabrication   IR LEDs are commonly fabricated to produce IR emissions at wavelengths of 850 nm, 880 nm, and 940 nm. Sizes available include 3 nm (T1), 5 nm (T1 – ¾), and surface mount technology (SMT) that results in a 3.4 nm by 8 nm diode size. Both the 3 nm and 5 nm LEDs are fabricated with two wires that represent the anode and cathode pin connections. Either LED can be used for applications that require "wire-wrap" and "through-hole." The SMT model is packaged in a plastic housing, which enables IR functionality to be incorporated into smaller electronic devices than a standard TV remote control.

5.1.10.2 Technical Details   To provide readers with an indication of the technical details associated with popular IR LEDs, this author examined the technical data provided by several vendor Web sites. The technical characteristics of both 3 nm and 5 nm IR LEDs popular offerings will be found in Table 5.2.

In examining the entries in Table 5.2, one of the more important ones for the consumer is the viewing angle. This is because a wider viewing angle permits the remote control user with more flexibility in using the device. That is, a wider emission angle allows the remote control user to point the control in the general direction of

Table 5.2   IR LED Characteristics

PARAMETER	SYMBOL	CONDITIONS	TYPICAL	MAXIMUM
Forward voltage	V_f	V_{cc} = Min, I_i = 12 mA	1.7 V	2.0 V
Reverse voltage	V_r	V_{cc} = Min, Ish = Max, V_{il} = mca		5.0 V
Maximum dc forward current	I_f	V_{cc} = Min, IoL = Max, V_{ih} = Max		250 mA
Peak current	I_p	V_{cc} = Min, IoL = Max, V_{ih} = Max		250 mA
Maximum power dissipation	P_d	V_{cc} = Max, V_i = 5.5 V		150 mW
Viewing angle		V_{cc} = Max, V_i = 2.4 V	20°	40°

the device to be controlled, whereas a lesser emission angle forces the remote control user to point the remote control either directly or much closer toward the device to be controlled.

5.1.10.3 Cost   Similar to other electronic components, the cost of IR LEDs has significantly decreased over the past half decade. After a bit of research, this author located pricing for IR LEDs during 2001. At that time, IR LEDs ranged in price from $30.00 to $300.00 each when purchased in a quantity of 1000 units, with the specific cost dependent on the size (3 mm or 5 mm) and technology (wired or SMT). In comparison, during 2008, the price of IR LEDs was between 16 and 28 cents per LED, with the cost based on the purchase of a single LED!

5.1.11 IR Detection with IR Photodiode

The most common device used to detect IR light is the photodiode. In this section, we will briefly discuss the operation and utilization of the IR photodiode.

Typically, the photodiode is used in conjunction with an optical filter that only passes wavelengths emitted by IR LEDs. Thus, the optical filter minimizes the potential of CFL, electric ballasts, and other light sources from interfering with the reception of IR signals.

5.1.11.1 Overview   A photodiode can be considered to represent an electronic light detector, which will convert light into either current or voltage. Fabricated as a p-n junction, when a photon of sufficient energy within a predefined wavelength strikes the diode, the result is an excitation of an electron, creating both a mobile electron and a positively charged electron hole. The resulting electron holes will move toward the anode, whereas the electrons will move toward the cathode, resulting in the generation of a photocurrent.

5.1.11.2 Modes   There are four common modes associated with photodiodes: photovoltaic, photoconductive, avalanche, and phototransistor. In the following four sections, we will note the purpose and the function of each.

5.1.11.2.1 Photovoltaic Mode   A photovoltaic-mode photodiode restricts the flow of photocurrent out of the device. This results in a voltage buildup. As the diode becomes forward biased, a "dark current," the constant response of the photodiode during periods when it is not exposed to light, begins to flow across the p-n junction in the opposite direction to the photocurrent. Another term used as a synonym for photovoltaic mode is *zero bias mode*.

5.1.11.2.2 Photoconductive Mode   In the photoconductive mode, the diode is usually reverse biased. When reverse biased, the width of the depletion region increases. This in turn decreases the capacitance at the junction, which reduces response time. Because the reverse bias causes only a small amount of current to be induced along its direction, this mode is both faster and normally exhibits less electronic noise. The current induced by the reverse bias is also referred to as *saturation* or *back current*.

5.1.11.2.3 Avalanche Photodiode   An avalanche photodiode is similar to a normal photodiode; however, the former is operated with a higher reverse bias. This action results in each photo-generated carrier being multiplied by the avalanche breakdown, which in turn results in an internal gain within the device. As an end result, the preceding increases its effective responsivity.

5.1.11.2.4 Phototransistor   A phototransistor represents a photodiode with internal gain. From a packaging point of view, the phototransistor represents a bipolar transistor that is enclosed within a transparent case so that light can reach the base–collector junction.

As light strikes the junction, electrons are generated. The resulting electrons are injected into the base, with the current then amplified by the transistor. Although phototransistors have a higher level of response to light than photodiodes, they are unable to detect low levels of light any better than a photodiode. In addition, a phototransistor has a slower response time than a photodiode.

5.1.11.3 Composition Previously, we briefly mentioned the p-n junction without discussing the material used in its creation. Currently, there are several types of semiconductor material used to create photodiodes. Table 5.3 lists four types of commonly used semiconductor materials, and their wavelength range in nanometers. Note that an IR photodiode would obviously use semiconductor material that can detect IR wavelengths.

Recall that visible light ranges from approximately 390 nm (violet) to 780 nm (red). Wavelengths from approximately 800 nm to 900 nm represent the so called "near" portion of IR light, which is used by a variety of remote control devices. Thus, from Table 5.3, any of the first three materials could be used to detect IR light. However, because silicon-based photodiodes have a significant band gap, they generate less noise than germanium-based photodiodes. Interestingly, germanium-based photodiodes can detect higher wavelengths. Thus, although both types of semiconductor material can be used in remote control detectors for applications beyond the near-IR range, germanium-based photodetectors are a more suitable semiconductor.

5.1.11.4 Packaging The packaging of a photodiode includes the previously mentioned optical filter. In addition, the basic photodiode will normally have a silicon nitrate layer placed over the diode, which functions as an antireflection coating. The plastic housing will more than likely use a dye as an optical filter. The dye will support transmission

Table 5.3 Semiconductor Materials Used In Photodiodes

SEMICONDUCTOR MATERIAL	WAVELENGTH RANGE (NM)
Silicon	190–1100
Germanium	400–1700
Indium gallium arsenide	800–2600
Lead sulfide	1000–3500

in the near-IR portion of the electromagnetic spectrum while strongly absorbing the visible light portion of the spectrum.

Depending on the manufacturer, there are different package sizes associated with photodiodes. In addition, it's important to examine both the method used to filter unwanted light as well as the wavelengths passed and filtered. This examination should obviously occur with respect to the intended characteristics of the IR transmitter to ensure compatibility between the emitter and detector.

5.1.11.5 Operation   The photodiode is used in a remote control application to generate a photocurrent on detection of an unfiltered IR signal. Figure 5.6 illustrates the use of a photodiode in a simple resistor-based detection circuit. In this example, the photodiode is connected in a reverse-biased manner across a power supply with a resistor (R) in series. When an applicable IR signal is detected, the photodiode will generate a current that will produce a voltage across the resistor. That voltage can then be amplified (A) to provide a required output level.

5.1.12 Selecting a Resistor

There are several constraints one must consider when selecting a resistor for the circuit previously shown in Figure 5.6. First, the smaller the value of the resistor, the smaller the voltage appearing across R. Conversely, as the value of the resistor is increased, the voltage appearing across R will increase. However, as we will soon note, there is a maximum value for the resistor based on the duration of the IR light pulses used in a remote control.

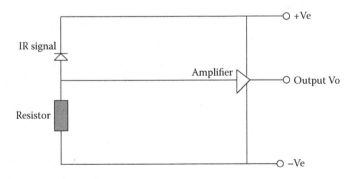

Figure 5.6   A simple photodiode detection circuit.

5.1.12.1 Limiting the Value of the Resistor As previously noted, as the value of the resistor decreases, the signal voltage drops. As the signal voltage drops, the gain associated with the amplifier (A) will increase. At a certain level, the amplifier gain will become a contributor to a significant level of noise that will make stability difficult to achieve. Thus, the value of the resistor needs to be above a certain value. Otherwise, noise will increase, which will reduce the range of the IR transmission that can be detected. Conversely, there is a maximum value for the resistor. This results from the fact that as the value of the resistor (R) increases, the cutoff frequency falls. Because many remote IR devices, such as a TV remote control, emit short-duration light pulses, the receiver circuit needs sufficient bandwidth to detect such pulses. As attenuation hinders the detection capability, another limitation that warrants attention is the selection of a maximum value for the resistor.

5.1.12.2 Maximum Resistance The resistance value commonly used in remote controls usually range between 100 KΩ and 300 KΩ, resulting in a circuit cutoff frequency between 300 kHz and 100 kHz. Although the use of a single resistor shown in Figure 5.6 works very well, problems can occur when the ambient lighting level significantly increases. In such situations, another possibly more stringent limitation is placed on the maximum value of the load resistor. This limitation results from the fact that a high level of ambient lighting will lead to a degree of photocurrent, which will cause a steady current to flow through the photodiode. That current causes a voltage across the resistor, which reduces the reverse bias applied to the photodiode. This in turn increases its capacitance, which reduces its bandwidth. As the value of the resistor (R) increases, a point will be reached where the voltage produced due to a high level of ambient light exceeds the supply voltage. When this occurs, the photodiode will become forward biased, and an IR signal resulting in photocurrent will be dissipated in its forward-biased shunt resistance. This in turn can severely attenuate the signal output. Thus, due to the previously mentioned problems, the use of a simple resistor in series with a photodiode is only useful when ambient light levels are low. For most home electronics such as televisions, DVD players, and CD players, the circuit shown in Figure 5.6 is usually sufficient for being used in their remote control units. The simplicity of the circuit results in low cost, which can be extremely important when a product run exceeds tens of thousands of devices. However, when ambient lighting reaches higher levels, more complex circuitry will be required. Now that we have an appreciation of the use of IR LEDs for remote control devices and the use of photodiodes to detect IR emission, let's turn our attention to another use of IR LEDs in communications. As we will shortly note, IR LEDs play a prominent role in several types of Ethernet local area networks (LANs).

5.2 Ethernet Networking

The original development of Ethernet LANs was at a data rate of 10 Mbps, using two different types of coaxial cable media. As Ethernet technology evolved and LANs became an indispensable part of corporate networks, a demand for faster data transmission was initially solved by the use of different types of unshielded twisted pair (UTP) wiring. However, the use of UTP resulted in a distance constraint. To expand transmission distance, several types of fiber-optic repeaters were developed, which operated first at 10 Mbps and later at the Fast Ethernet data rate of 100 Mbps. Although such repeaters successfully extended the range of Ethernet networks, it was the development of Gigabit and 10 Gigabit networks that could use fiber as a transmission medium which provided both the high data rate and extended transmission distance required by many organizations. Although LEDs have been successfully used for several decades to extend the range of Ethernet and Fast Ethernet networks, unfortunately their maximum data rate with existing state-of-the-art technology is limited to 622 Mbps. Thus, although Gigabit and 10 Gigabit Ethernet networks cannot be driven through the use of LEDs at the present time, they represent a viable mechanism for extending the reach of other types of Ethernet networks.

In Section 5.2.1, we will commence our examination of the use of LEDs in Ethernet networks by briefly discussing how fiber-optic cable works. Once this is accomplished, we will describe and discuss the use of LEDs in certain types of Ethernet networking environments.

5.2.1 Fiber-Optic Cable

Fiber-optic cable is used to transmit data in the form of light generated by an LED or a laser diode. As light in the form of photons traverses the cable, its strength decreases as the distance increases. This decrease is referred to as *attenuation* and measured in decibels (dB). On a typical fiber-optic cable, the attenuation or power loss is 3 dB/km.

5.2.1.1 Decibels Power Measurements

The decibel (dB) is a standard measurement to compute power gain or loss. Mathematically,

$$dB = 10 \log_{10} \frac{\text{output power}}{\text{input power}}$$

To illustrate the use of the decibel, it's important to remember two common relationships. First, the log of $1/x$ equals a negative log of x, or mathematically:

$$\text{Log } 1/x = -\log x$$

The second relationship is the fact that the log to the base 10 of x is the same as determining what power 10 has to be raised to, to equal the value of x.

To illustrate the preceding relationships, assume the output power is one-tenth of the input power. Then, the loss in dB becomes:

$$dB = \frac{1}{10 \log_{10} \dfrac{10}{1}} = 10 \log_{10} \frac{1}{10}$$

$$= -10 \log_{10} 10$$

$$= -10$$

Now let's assume the output power is one-hundredth the input power. Thus,

$$dB = \frac{1}{10 \log_{10} \dfrac{100}{1}}$$

$$dB = -10 \log_{10} 100$$

$$dB = -10.2 = -20$$

Table 5.4 The Power Out As A Percentage of Power In and the Percentage Power Loss for 20 dB

dB	POWER OUT AS A PERCENTAGE OF POWER IN	PERCENTAGE OF POWER LOSS
0.00	100.0	0.0
0.001	99.98	0.02
0.01	99.8	0.22
0.05	99.0	1.0
0.10	98.0	2.0
0.20	95.5	4.5
0.30	93.0	7.0
0.40	91.0	9.0
0.50	89.0	11.0
0.60	87.0	13.0
0.70	85.0	15.0
0.80	83.0	17.0
0.90	81.0	19.0
1.00	79.0	21.0
2.00	63.0	37.0
3.00	50.0	50.0
4.00	40.0	60.0
5.00	32.0	68.0
10.00	10.0	90.0
20.00	1.00	99.0
30.00	0.10	99.9

Note that the $\log_{10} 10$ is the same as determining that 10 raised to the power of 2 is 100. From the preceding equations, we can note that the smaller the output with respect to a fixed input, the more negative the dB or power loss. Conversely, if we placed an amplifier in the cable so that the output power was greater than the input power, the dB value would become positive and would increase as the ratio of output to input power. When the dB power loss is zero, this special case indicates that the output and input power levels are the same, thus there is no gain or loss.

Table 5.4 provides readers with the power out as a percentage of power in and the percentage power loss for 20 dB.

5.2.1.2 Single versus Dual Cables Initially, two cables were required for an Ethernet connection using light. One cable functioned as the transmitter, which was terminated with a light-sensitive receiver at

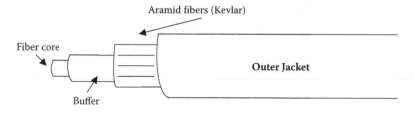

Figure 5.7 The composition of a fiber-optic cable.

the other end of the cable. The second cable functions in reverse, enabling full-duplex communications to be supported. Recently, the use of two distinct wavelengths enables transmission and reception of data to occur over a single fiber-optic cable.

5.2.1.3 Cable Composition Figure 5.7 illustrates the composition of a fiber-optic cable. Light is transmitted through the core of the cable. The cladding surrounds the core. Its internal mirrorlike reflectivity keeps light flowing inside the core and minimizes power loss. Surrounding the cladding is a buffer consisting of a hard plastic coating, which protects both the core and cladding. Next, the plastic coating is surrounded by aramid fibers, which provide the strength that enables the cable to be pulled through conduits, along trenches, and through similar constricted locations. Often the aramid fibers are manufactured by DuPont Corporation and are referred to as *Kevlar®*. Finally, surrounding the fibers is an outer jacket consisting of PVC plastic or a flame-retardant compound.

5.2.1.4 Types of Fiber Cable There are two basic types of fiber-optic cables named according to the manner in which light flows in the cable: single mode and multimode. A single-mode cable has a very thin core, typically 9 nm. Light flows on a single path through this type of cable. Due to the thin core, the light source or transmitter is a laser diode. In comparison, a multimode cable has a much larger core, typically today either 50 nm or 65 nm, and supports the transmission of light along multiple paths as it enters the core in several directions. Because LEDs transmit light in several directions, they are well suited for use with multimode fiber.

5.2.1.5 Fiber and Wavelength Light is transmitted via fiber-optic cable in various wavelengths, with the wavelength used depending on the type of fiber (single mode or multimode) and the transmission distance. Common wavelengths associated with single-mode fiber are 1310 and 1550 nm, whereas wavelengths of 850 and 1300 nm are used with multimode fiber. Note that, for both single-mode and multimode fiber cable, the wavelengths used for transmission are in the IR region and are thus invisible to the human eye because we can only see colors ranging between violet at approximately 400 nm and deep red at 700 nm.

The International Commission of Illumination (CIE) recommended the division of IR radiation into three bands. These bands are listed in Table 5.5.

It's interesting to note that the multimode wavelengths fall into what is considered to represent the near-IR region, whereas single-mode wavelengths fall into that region and the short-wavelength IR region.

There are two types of multimode fiber cables commonly used in LANs that differ depending on their core diameter. They are 50 micron (μm) and 62.5 μm diameter fibers. Until the turn of the century, fiber-optic cable with a core diameter of 62.5 μm was used in most LANs. The 62.5 μm cable has a lower modal bandwidth than 50 μm diameter cable, which affects its ability to transmit light. However, although its* diameter is larger, its manufacture is less expensive, which results in a lower cost to the consumer.

The use of 62.5 μm diameter fiber cable was sufficient for use at Ethernet 10 Mbps data rate and the 100 Mbps data rate of Fast Ethernet. With the development of Gigabit and 10 Gigabit Ethernet LANs, more bandwidth was required. Because 50 μm multimode has approximately three times the bandwidth of 62.5 μm fiber, this allows the smaller-diameter fiber to be used to transmit longer distances. Today, many organizations using Gigabit Ethernet employ

Table 5.5 CIE IR Band Recommendations

BAND	WAVELENGTH (NM)
IR-A	700–1,400
IR-B	1400–3,000
IR-C	3000–1,000,000

both 62.5 and 50 μm diameter fibers. The 62.5 μm fiber supports the use of LEDs. It's commonly used in relatively short cable runs such as vertical rises or to support fiber-based Ethernet and Fast Ethernet. In comparison, 50 μm fiber can be used with either LEDs or relatively newly developed 850 nm vertical-cavity surface-emitting lasers (VCSELs). Although you can use 62.5 μm fiber to achieve Gigabit and 10 Gigabit data rates, 50 μm fiber has a lower level of attenuation that enables an extended transmission distance to be achieved. As we will shortly note later in this chapter, although the maximum data rate obtainable through the use of LEDs is 633 Mbps, they currently cannot be used to support Gigabit and 10 Gigabit networks. However, they are currently used to support many optical carrier (OC) networks that provide optical transmission from communications carrier offices into a customer premises to include their use in Synchronous Optical Network (SONET) rings that provide a high degree of communications redundancy.

Typically, the selection of 62.5 μm versus 50 μm fiber depends on the data rate required, distance, upgradeability, and economics. When installing a new network or network segment, most organizations today will probably elect to use 50 μm fiber. If extending an existing 62.5 μm fiber-based network, organizations are likely to opt for using the same fiber as this will usually result in a better level of performance than when the two types of multimode fiber are mixed. Now that we have an appreciation for the different types of multimode fiber, let's turn our attention to examining the different types of Ethernet LANs that can use fiber cable. In doing so, we will literally start at the beginning by turning our attention to 10 Mps operations. Once this is accomplished, we will examine 100 Mbps, Gigabit, and then 10 Gigabit operations.

5.2.2 FOIRL and 10BASE-F

The first use of fiber-optic technology with Ethernet was to extend the transmission range of 10 Mbps LANs, referred to as the *fiber-optic repeater link* (FOIRL). This technology represents one of two fiber-optic specifications that were developed to support 10 Mbps Ethernet networks. FOIRL was later replaced by a second standard referred to as *10BASE-F*, which provides backward compatibility with the older specification.

5.2.2.1 Overview   Under the original FOIRL standard, a fiber link could traverse up to 1000 m between two repeaters. In comparison, the 10BASE-F standard doubles the distance to 2000 m. In addition to the doubling of the range of transmission, other differences between the two specifications include the names of different fiber-optic components associated with each networking technology and such optical specifications as the transmit power levels, receiver sensitivity, and power loss or attenuation. Because the optical transceiver is common to both standards, let's commence our examination of both network range extenders by turning our attention to this device.

5.2.2.2 Optical Transceiver   As previously noted, the optical transceiver is common to both the FOIRL and 10BASE-F specifications, with only transmit and receive optical power levels differing between the two. An optical transceiver consists of a pulse-generating IR LED, a photodetector, and associated transmit and receive circuitry. The transmit circuitry turns the IR LED on and off to convert electrical voltage representing data into a series of light pulses for transmission over the fiber. As previously noted, the photodetector recognizes light pulses and generates a current that provides an amplified voltage. The receive circuitry generates and shapes electrical pulses to correspond to the received light pulses.

Optical transceivers can be obtained as stand-alone units mounted on an adapter card for insertion into the system unit of a computer or built into a fiber hub. Thus, our next step is to discuss the fiber hub.

5.2.2.3 The Fiber Hub   Under the FOIRL specification, the use of a fiber-optic segment was only applicable between repeaters. This meant that you could not directly connect a distant computer to a port on a hub. Recognizing this requirement for connectivity resulted in the Institute of Electrical and Electronic Engineers (IEEE) promulgating the 10BASE-F standard.

A fiber hub represents a special type of hub that contains a number of FOIRL ports; one attachment unit interface (AUI), a 15-pin port for connecting the hub to a transceiver copper cable, which in turn can be connected to a coaxial-cable-based network; and usually one or more 10BASE-T, twisted-wire-based devices. You can use a fiber hub to support one or more extended distance Ethernet connections, linking those connections directly to a 10BASE-T network by using a 10BASE-T port built into the fiber hub or indirectly to another type of older Ethernet network via the AUI port. If we focus on the more modern 10BASE-F hub, we would note that it consists of a series of fiber link (10BASE-FL) ports, with the main difference between 10BASE-F and FOIRL hubs residing in their optical transmit power and receiver sensitivity supported by the ports on each hub.

5.2.2.4 Fiber Adapter A third hardware component used with FOIRL is a fiber adapter. It can be considered to represent a media conversion device. The fiber adapter converts electrical signals into optical signals and vice versa. The primary purpose of the fiber adapter is to extend the transmission distance between a wire hub and either an attached workstation or another 10BASE-T wire hub. Through its use, transmission can be extended from 100 m (328 ft) to 1000 m. When a 10BASE-F fiber adapter is used with a 10BASE-F fiber hub, the transmission distance can be extended to 2000 m. For both FOIRL and 10BASE-F, unless the fiber adapter is connected directly to a fiber hub, an adapter is required at each end of an extended fiber link.

5.2.2.5 Wire and Fiber Distance Limitations To obtain an appreciation of distance limitations, Figure 5.8 illustrates the use of FOIRL and 10BASE-T compliant fiber adapters and hubs. Note that when

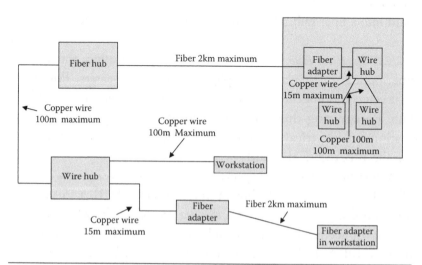

Figure 5.8 FOIRL and 10BASE-T transmission limits.

10BASE-T is used with FOIRL compliant fiber hubs and fiber adapters, the maximum transmission distance depends on the devices at the end points. For example, transmission between a fiber hub and a wire hub performed via UTP is limited to a maximum distance of 300 m.

In examining Figure 5.8, it was assumed that three 10BASE-T wiring hubs were installed in a wiring closet in one building. In another building located approximately 2000 m away, users are serviced by an additional wire hub whose connectivity is via a connection to a fiber hub. A stand-alone PC or workstation, as computers connected to a LAN are referred to, is located in a third building. Through the use of FOIRL transmission, all three locations can be interconnected into a common network.

5.2.3 10BASE-F

The 10BASE-F standard represents the first official use of LED technology standardized for LANs. 10BASE-F includes three types of segments for optical transmission: 10BASE-FL, 10BASE-FB, and 10BASE-FP.

5.2.3.1 10BASE-FL

The 10BASE-FL standard provides a doubling of the range of FOIRL transmission, supporting a fiber-optic link up to 2000 m in length as long as 10BASE-FL hardware is used at both ends of a segment. Otherwise, mixing of 10BASE-F and FOIRL hardware will reduce the maximum length of the optical segment to 1000 m.

10BASE-F was developed as a replacement for FOIRL. Unlike FOIRL, which was restricted to providing an optical connection between repeaters, 10BASE-T supports an optical segment to be used between two workstations, two repeaters, or between a workstation and a repeater port.

5.2.3.1.1 Connection Methods

There are two methods by which a copper-based network node can be connected to a 10BASE-FL segment. In the first method, a stand-alone fiber-optic Media Attachment Unit (MAU), commonly referred to as a FOMAU, can be connected through the use of a 15-pin AUI connector on a network adapter to obtain an electrical-to-optical conversion capability. In the second

Workstation (PC) system unit

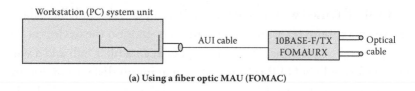

(a) Using a fiber optic MAU (FOMAC)

Workstation (PC) system unit

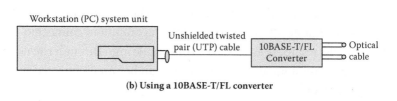

(b) Using a 10BASE-T/FL converter

Figure 5.9 Options for connecting to a 100BASE-FL segment using a fiber-optic MAU (FOMAU).

method, a 10BASE-T/FL converter is used. The converter is used only when a 10BASE-T port is available on a network adapter. Both the 10BASE-FL FOMAU and the 10BASE-T/FL converter include two fiber-optic connectors, one for transmitting and one for receiving data. Some 10BASE-T/FL converters can be obtained with two RJ-45 connectors, which increases cabling flexibility. One RJ-45 connector functions as a crossover cable that supports hub-to-hub communications, whereas the second functions as a straight-through connection. If you turn your attention to the top portion of Figure 5.9, you will note a computer or workstation connected to a 10BASE-FL FOMAU via a 15-pin AUI connector. In the lower portion of the previously referenced figure, the use of a 10BASE-T/FL converter is shown. The use of either device provides you with the ability to transmit up to 2000 m via a fiber link when 10BASE-F compliant hardware is used at both ends of an optical link.

5.2.3.1.2 10BASE-FL Hub A 10BASE-F optical hub is obtained by the inclusion of two or more FOMAUs in the hub. This results in the hub in essence functioning as an optical repeater, retransmitting data received on one port to all other ports similar to the manner in which conventional copper-wired hubs operate. Under the 10BASE-F standard, you can use multiple 10BASE-FL connections to link several individual LAN stations at distances up to 2000 m to a common hub equipped with FOMAU ports.

5.2.3.2 10BASE-FB 10BASE-FB represents a second 10BASE-F specification, with 'B' used to indicate a synchronous signaling backbone segment. The 10BASE-FB specification was developed to overcome the limit on repeaters due to the "Ethernet 5-4-3" rule, according to which no two stations can be separated by more than five segments connected by four repeaters or hubs, of which only three can contain user connections, a term referred to as *population*. Under this rule, the wiring that joins each hub together is considered to be a segment, whereas each hub is considered to be a repeater. A 10BASE-FB signaling repeater can be used to connect repeater hubs together into a repeated backbone network that can span multiple 2000 m links.

5.2.3.3 10BASE-FP The third 10BASE-F specification, referred to as *10BASE-FP*, was developed to support the connection of multiple stations via a common segment that can be up to 500 m in length. The 'P' in 10BASE-FP references the fact that the end segment is a fiber-passive system. The 10BASE-FP specification permits a single fiber-optic passive star coupler to be used to connect up to 33 stations, with each station located up to 500 m away from a hub, via the use of a shared fiber segment. Recall that a passive fiber-optic network has no electrical input to add energy to the information signal. Instead, splitters are used to distribute a single fiber's capacity among multiple fiber strands.

5.2.4 Optical Media Support

Both 50 μm and 62.5 μm multimode fibers can be used with 10BASE-F hardware. Table 5.6 provides a comparison of optical attenuation for 10BASE-FL and FOIRL for six types of multimode fiber, of which 50/125 and 62.5/125 are by far the most commonly used. Recall that the first digit references the core diameter in microns, whereas the

Table 5.6 Comparing the Optical Loss Budget (Attenuation) of 10BASE-FL and FOIRL

Multimode graded Index fiber size (μm)	50/125	50/125	50/125	62/125	83/125	100/140
Numerical aperture	0.20	0.21	0.22	0.275	0.26	0.30
10BASE-FL loss budget (dB)	9.7	9.2	9.6	13.5	15.7	19.0
FOIRL loss budget (dB)	6.7	7.2	7.1	11.0	13.2	16.5

second digit references the diameter of the cladding in microns. LEDs used for transmission have a wavelength of 850 nm, and their optical loss budget ranges between 9.7 and 19.0 dB, with the exact amount dependent on the type of fiber used.

5.2.5 Fast Ethernet

There are two Fast Ethernet fiber-optic standards that support a data transmission rate of 100 Mbps and support the use of LEDs. Those standards are 100BASE-FX and 100BASE-SX. Similar to 10BASE-T, which operates at 10 Mbps, the various versions of Fast Ethernet that operate at 100 Mbps over copper wire are restricted to a maximum network segment distance of 100 m or 330 ft. Thus, the two versions of Fast Ethernet for transmission over optical fiber provide a mechanism to extend the transmission distance similar to the manner in which 10BASE-FL extends the range of 10 Mbps 10BASE-T network.

5.2.5.1 100BASE-FX The first optical fiber standard developed for Fast Ethernet is 100BASE-FX. This standard defines the use of a 1300 nm near-IR light wavelength transmitted via dual optical fiber, with one fiber for transmission and the other for reception. Several vendors market 100BASE-FX devices that use 1310 nm LEDs which operate on 50 or 62.5 µm multimode fiber. To ensure that collisions are detected, the maximum transmission distance is limited to 400 m (1310 ft) for half-duplex and 2 km (6600 ft) for full-duplex operation. To obtain an extended transmission distance, single-mode optical fiber can be used with a 1310 nm laser. Similar to the use of optical fiber at 10 Mbps, the use of 100BASE-FX can extend a network segment or individual workstations to a hub equipped with 100BASE-FX ports. At the workstation, a variety of 100BASE-FX network adapter cards can be obtained, which when installed enables individual workstations to be linked to a hub.

5.2.5.2 100BASE-SX A second version of Fast Ethernet over optical fiber is the 100BASE-SX specification. This Fast Ethernet over optical fiber standard represents a low-cost alternative to 100BASE-FX as it uses short-wavelength optics, which is less expensive than the long-wavelength (1310 nm) optics used in 100BASE-FX. Specifically, 100BASE-SX uses an 850 nm LED. Because of the shorter wavelength, the maximum transmission distance of 100BASE-SX is reduced to 300 m (980 ft).

5.2.6 *Gigabit Ethernet*

Gigabit Ethernet represents a logical progression of the data rate of Ethernet through Fast Ethernet. That is, Gigabit Ethernet's 1Gbps data rate is ten times that of Fast Ethernet, which was ten times that of Ethernet. To overcome the challenges associated with moving from 100 Mbps to 1Gbps, two technologies were in effect merged at the physical layer to provide a 1 Gbps data rate. Those technologies were the IEEE's 802.3 Ethernet standard and the American National Standards Institute (ANSI) Fibre Channel Standard. Through combining portions of each technology, it became possible to take advantage of the high-speed physical interface characteristics of the Fibre Channel while maintaining the IEEE 802.3 Ethernet frame format. The new Gigabit standard is known as the *IEEE's 802.3z Gigabit Ethernet standard.*

Although LEDs were used to support Ethernet and Fast Ethernet over optical fiber, the development of Gigabit Ethernet and later 10 Gigabit Ethernet created problems that precluded the use of LEDs.

LEDs transmit a signal over a relatively wide spectral bandwidth. Hence, they are well suited for use with multimode fiber. Because multimode fiber has a graded refractive index, this results in light traveling at different speeds in the fiber, faster toward the interface between the core and the surrounding and slower at the center of the core. Because a graded index fiber is not completely uniform, this means that some signals arrive at the opposite end of the fiber before other signals, creating what is referred to as *modal dispersion*. Owing to modal dispersion as well as power loss and the inability to turn on and off quickly enough to support higher bandwidth, the maximum data rate that can be obtained through the use of LEDs and standard multimode fiber is 662 Mbps. This limitation resulted in the use of vertical-cavity surface-emitting lasers (VCSELs) operating at 850 nm to provide the higher data transfer capability required by Gigabit

Ethernet. It should also be noted that VCSELs have much faster rise and fall times than LEDs as well as more power and a smaller light illumination. Because a larger fiber core enables more modes of light to traverse, the fiber and VCSELs can cause a bit of mode delay. Industry has moved from the use of 62.5 μm to 50 μm fiber for Gigabit Ethernet. For 10 Gigabit Ethernet, an enhanced "laser-optimized 50 μm multimode fiber" is now being used with VCSELs to support a transmission distance of 300 m at 10 Gbps.

From an LED perspective, it's important to note that their use in communications applications reached a limit. That limit, which is based on the modal dispersion of LED light as well as the rise and fall times associated with turning the device on and off, results in a maximum obtainable data rate of 622 Mbps with current technology. Although not sufficient for supporting Gigabit or 10Gigabit Ethernet, LEDs are commonly used for optical carrier (OC) long-distance transmission at data rates of 155.52 Mbps (OC-3) and 622.08 Mbps (OC-12).

6
COMPARING LEDs
AND LASER DIODES

Although the primary emphasis of this book is on light-emitting diodes (LEDs), this author would be remiss if he did not address the laser diode. In this chapter, we will describe and discuss the meaning of the term *laser diode*, including its fabrication and operation. Because the generation of coherent light is one of the key characteristics of lasers, we will also examine what this term means, which will serve as a refresher for some readers. Once this is accomplished, we will turn our attention to several types of laser diodes and how they supplement the use of LEDs in certain applications as well as enable other applications that cannot be supported by the LED.

As we describe and discuss the laser diode, we will periodically compare and contrast it with the LED. Because both devices are based on the use of similar semiconductor materials but operate differently, in the second section of this chapter, we will compare and contrast their operation. In doing so, we will describe and discuss several key differences between the laser diode and the LED, which include but are not limited to areas such as their focus, wavelength, power, safety, and other parameters, to make readers aware of the major differences between the two devices. This in turn will provide readers with knowledge concerning some of the applications each device is well suited to perform.

6.1 The Laser Diode

A laser diode is a coherent light emitter that although similar to an LED has some significant differences from that device. As we investigate the laser diode in this chapter, we will note that there are many different types of such devices. The most common laser diode is similar to an LED in that it is formed from a p-n junction, with crystal being doped to produce an n-type region and a p-type region, one above the other. Similar to an LED, a forward electrical bias results in holes and electrons being injected from opposite sides of the p-n junction into the depletion region. Like the LED, holes are injected from the doped p-type region, whereas electrons are injected from the doped n-type region.

6.1.1 Emission of Coherent Light by Laser Diodes

If you ask a person the difference between lasers and LEDs, one of the first operational characteristics that might be mentioned is that lasers generate coherent light whereas LEDs do not. However, if you probe a bit further and ask how coherent light is produced, you might receive a puzzled look. Thus, to remove that potential puzzled look, let's turn our attention to one of the unique properties of lasers, which is the production of coherent light.

If we remember that the term *laser* is an acronym for "light amplification by stimulated emission of radiation," then any discussion of coherent light needs to first focus on the term *stimulated emission*. Because that term represents a quantum process, we need to back up a bit and start at the beginning. To do so, let's first discuss the quantum process.

6.1.1.1 The Quantum Process As we noted earlier in this book, the energy levels of atoms and molecules can only have certain quantized values. As current flows through certain doped semiconductor materials, the energy levels can change. This change in an energy level can result in absorption or emission of photon energy.

In the later 1900s, the now well-known German scientist Max Planck assumed that radiant energy was emitted in small bursts that were referred to as *quanta*. Each of the bursts known as a *quantum* has a level of energy (E), which depends on the frequency (f) of the electromagnetic radiation such that

$$E = h \times f$$

where h is the Planck's constant, which is 6.62618×10^{-34} J/s. Based on Planck's equation, energy is proportional to the frequency of radiation.

Returning to the quantum process, the transition of photons between states in a semiconductor can result in absorption, emission, or stimulated emission. Thus, let's turn our attention to each possibility.

6.1.1.1.1 Absorption and Emission   The absorption of a photon generated by an electrical transition will only occur when its quantum energy level matches the energy gap between the initial and final energy states. That is, if we assume the energy gap is $E_2 - E_1$, then absorption will only occur when the change in the energy $\delta E = E_2 - E_1$.

The top portion of Figure 6.1 illustrates the absorption process, whereas the lower portion illustrates the emission of a photon of energy. Note that absorption occurs only when the quantum energy of the photon matches the energy gap, whereas a downward transition represents the emission of a photon of energy.

6.1.1.1.2 Stimulated Emission   Now that we have an appreciation for absorption and emission, let's focus our attention on stimulated emission. For stimulated emission to occur, an electron will be in an upper energy level or excited state. Then, an incoming photon whose quantum energy is equal to the difference between the electron's present level and a lower level can stimulate a transition of the electron from the excited state to the lower level, resulting in the generation of a second photon that has the same amount of energy. By applying a forward current on a correctly doped semiconductor, a sizable population

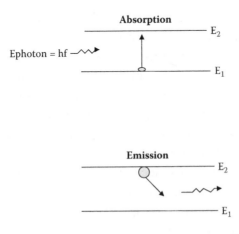

Figure 6.1   Absorption and emission of photons.

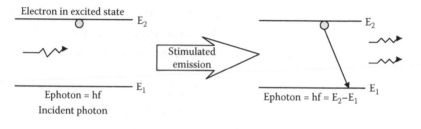

Figure 6.2 The stimulated emission process.

of electrons will reside at an upper energy level. When this situation occurs, the condition is referred to as *population inversion* and is the foundation for the stimulated emission of multiple photons. Figure 6.2 illustrates the stimulated emission process. Note that this process represents the precondition for the light amplification process that occurs in a laser, which will shortly be described when we discuss the use of mirrors. Returning our attention to Figure 6.2, because the emitted photons have a defined time and phase relationship to one another, this results in emitted light having a high degree of coherence.

The result of a significant population inversion is a large number of electrons at a high-energy state, which is shown as E_2 in Figure 6.2. When a significant population inversion exists, it becomes possible for stimulated emission to produce significant light emission in the form of photons that have a definite phase relationship. This phase relationship produces coherent light. Because a laser produces light that has a single wavelength, we can also say laser light is monochromatic.

6.1.1.2 Use of Mirrors   Although we will shortly discuss the semiconductor materials used for the creation of laser diodes, let's turn our attention to the use of mirrors that results in the light amplification process, which represents the first two terms in the acronym laser. Figure 6.3 illustrates how lasers use two mirrors placed on opposite sides of the material used to stimulate photon emission. One mirror is 100 percent or fully reflective, whereas the mirror at the opposite end is approximately 99 percent reflective. As photons are emitted, they are fully reflected from one side of the laser and mostly reflected from the opposite end. However, as the 99 percent reflective mirror allows 1 percent of the photons to escape, they exit the laser in the form of coherent, monochromatic light.

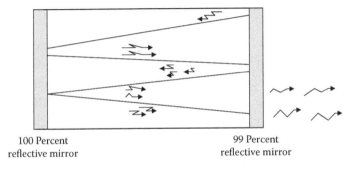

100 Percent   99 Percent
reflective mirror   reflective mirror

Figure 6.3   The use of mirrors to provide light amplification.

6.1.2 Reviewing LED and Laser Diode Operations

In an LED, the energy released from the recombination of electrons and holes results in the emission of light via the generation of photons. The emission of photons generates incoherent light due to the flow of photons in different directions. The result of this action is that LEDs generate multiple modes of light. In comparison, through the use of a different type of semiconductor material, a laser diode generates coherent light. In actuality, it is the energy level differences between the conduction and valence band electrons in the semiconductor material that enables the laser diode to generate coherent light.

6.1.3 Evolution of Laser Diodes

The first demonstration of coherent light emission from a semiconductor diode was performed by Robert Hall and his team of fellow scientists at a General Electric Corporation research center during 1962. This demonstration was followed by the development of a visible-wavelength laser diode that was presented by Nick Holonyak, Jr., later in 1962.

During the 1960s and 1970s, laser diodes evolved considerably from homojunction devices to heterojunction diode laser devices. By the mid-2000s, almost a billion diode lasers were sold, being embedded in a variety of applications, and several new types of laser diodes were developed, including vertical-cavity surface-emitting lasers (VCSELs), which will be described in Section 6.1.4.4.

6.1.4 Types of Laser Diodes

Since 1962, a variety of laser diodes have been developed. Some of the different types of laser diodes are edge-emitting laser diodes, double heterostructure lasers, quantum well lasers, distributed feedback lasers, and VCSELs. Because the edge-emitting laser diode is not only a commonly used laser but its operation also forms the basis for the creation of other types of lasers, we will first examine its fabrication and operation. Using this laser as a base, we will briefly discuss the next three types of lasers and then focus our attention on examining in detail the operation of vertical external cavity surface-emitting lasers (VECSELs).

6.1.4.1 Edge-Emitting Laser Diode   One of the more common types of laser diodes is the edge-emitting laser, which is illustrated in Figure 6.4. In an edge-emitter laser diode, the light exits from the edge of the laser diode chip. This type of laser falls into a category of lasers referred to as *Fabry–Perot* laser diodes, as its cavity is very similar to that in a conventional gas or solid-state laser, but formed inside the semiconductor laser diode chip. As we will shortly note, mirrors are used at each edge of the chip with different levels of antireflection coating to provide amplification of light.

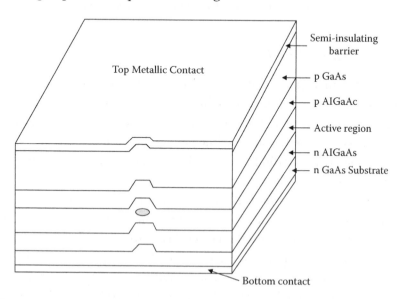

Figure 6.4   Edge-emitting laser diode.

In examining Figure 6.4, note that coherent light in the form of a laser beam would exit from both ends of a crystal formed with p-type and n-type semiconductors. To prevent this, mirrors can be used at both ends of the crystal, or the ends could be optically ground, polished, and coated. The mirror or the ground and polished end would have 100 percent reflection, whereas the opposite end would be only 99 percent reflective, resulting in light amplification.

When there is only one p-n junction, the laser diode configuration is referred to as a *homojunction* diode. Because there are several benefits to the use of two or more closely spaced junctions formed by layers of p-type and n-type material, certain types of laser diodes are fabricated in this manner. Such lasers are referred to as *heterojunction* laser diodes.

The modification of the structure of the material used to create a laser diode resulted in the development of a variety of different types of devices. One, which we will discuss at length, is the vertical-cavity surface-emitting laser (VCSEL) diode, which is used to provide high-speed communications over fiber-optic cable that is beyond the capability of LEDs (Section 6.1.4.4).

Returning to Figure 6.4, current will flow from the p-type to the n-type semiconductor material, as in an LED, with electrons and holes being injected into the active region. Because the figure illustrates the use of dual p-type and n-type semiconductor layers, the resulting laser diode has a double heterostructure configuration. This results in the band gap of the semiconductor having a lower refractive index than that in the active region. Thus, the emitted photons are guided into a coherent light pattern.

6.1.4.1.1 Size and Power   Two of the key attributes of laser diodes are their size and power. Concerning size, laser diodes are extremely compact, with the active element approximately the size of a grain of sand. With respect to power, laser diodes operate using low voltage and are considered to be extremely efficient. Some laser diodes have an electrical-to-optical efficiency greater than 50 percent such that you obtain 1 W of light for every 2 W of dc power provided. In comparison, as noted previously in this book, the efficiency of incandescent lighting is at best a few percent as most of the power applied to this type of lighting source is converted into heat.

6.1.4.1.2 Common Applications   In a modern environment, the laser diode has become as ubiquitous as the personal computer. If you drive a modern vehicle, jog using a CD player to listen to the latest songs or golden oldies, chances are very high that you are using a laser diode. When you visit a pharmacy and have your purchases scanned or burn a new set of songs on the CD-ROM drive installed in your personal computer, once again the laser diode provides the functionality that enables the application to be performed. Because they can be quickly turned on and off, the laser diode has been used in communications, especially when data rates need to exceed the 622 Mbps limitation of LEDs. Thus, laser diodes are used in Gigabit and 10 Gigabit Ethernet networks. In addition, when you attend a lecture, chances are very high that the instructor will use either a green- or red-laser-diode-based pointer.

Because laser diodes generate a narrow beam of light, they can be easily used for applications that require directed energy of an optical beam. Such applications can include range finders on tanks, where the time for the reflection of the beam can provide a range to target, to bar code readers and the handheld scanner in big-box stores that enable a clerk to scan the UPC on large items too heavy or bulky to be lifted onto a checkout counter. Thus, the laser diode has rapidly moved from the laboratory into numerous applications that now govern a good portion of our daily activities.

6.1.4.2 Double Heterostructure Laser   A double heterostructure (DH) laser is created by the placement of a low-band-gap material between two high-band-gap layers. Typically, gallium arsenide (GaAs) is used with aluminum gallium arsenide (Al$_x$GaAs). Because each of the junctions between different band gap materials represents a hetero-structure, this results in a double heterostructure (DH) laser.

By confining the active region, where free holes and electrons simul-taneously exist, to a thin middle layer, this allows more electron–hole pairs to contribute to the amplification process. In addition, because light is reflected from the heterojunction, this confines photons to the region where amplification occurs, boosting the output of the laser.

6.1.4.3 Quantum Well Laser   Another type of laser diode is fabri-cated by making the middle layer sufficiently thin so that it acts as a

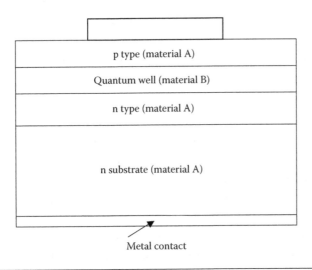

Figure 6.5   A quantum well laser diode.

quantum well, in effect quantizing the vertical variations of an electron's wavefunction. This increases the efficiency of the laser.

Figure 6.5 illustrates a front view of a quantum well laser diode.
Note that a laser diode that contains multiple quantum well layers is
referred to, as you might expect, *as a multiple quantum well laser.* Such
lasers have an increase in efficiency over a quantum well laser.

6.1.4.4 Vertical-Cavity Surface-Emitting Laser (VCSEL)   Another type
of laser diode that warrants attention is the VCSEL. As its name
implies, a VCSEL emits laser (coherent) light vertically from its surface and is fabricated so that it has a vertical laser cavity. That cavity
is actually formed by the fabrication of numerous layers of mirrors
and active laser semiconductor materials, all formed epitaxially on an
inactive substrate.

The VCSEL has the optical cavity axis along the direction of flow
of current. In comparison, the flow of current is perpendicular to the
optical cavity axis in conventional laser diodes.

The major difference between a conventional edge-emitting laser
diode and the VCSEL in addition to the direction of light flow (vertical versus edge) is their fabrication. In a VCSEL, the active region,
which is an electrically pumped gain region, emits light. The placing of layers of different semiconductor materials above and below
the gain region results in the creation of mirrors that reflect a range

of wavelengths back into the cavity, which results in the emission of light at a single wavelength.

The difference in the fabrication of a VCSEL from a conventional edge-emitting laser diode results from the ability to easily shape the emitting region for optimum use in an application. This can be extremely important as the shape of the emitting region can be tailored, for example, for coupling to a particular type of optical fiber. In comparison, other types of laser diodes are typically long and narrow because the junction where laser action occurs has a minimum thickness and thus cannot have their emitting regions shaped.

Figure 6.6 illustrates the fabrication of a VCSEL laser diode. Note that the active region's length is relatively short in comparison to its lateral dimensions, enabling radiation to exit from the surface of the cavity instead of from its edge. The reflectors, which are positioned at the ends of the cavity, are dielectric mirrors. Such mirrors consist of alternating high and low refractive p- and n-type materials.

The actual VCSEL structure is quite complex, with Bragg reflectors having as many as 120 mirror layers to provide the reflection required. The fabrication process results in high-efficiency mirrors that produce a relatively low threshold current, typically below 1 mA.

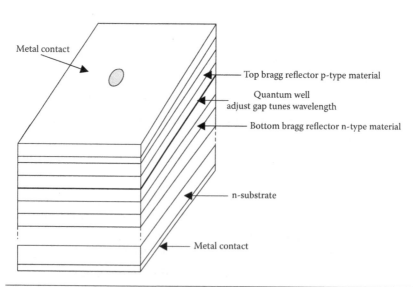

Figure 6.6 The structure of a vertical cavity surface-emitting laser (VCSEL).

6.1.4.5 Trade-offs between Various Laser Diodes   There are certain trade-offs associated with the use of VCSELs, edge emitters, and other types of laser diodes that you need to consider. Because edge-emitters cannot be tested until the end of the production process, it's possible to waste both materials and production time producing laser diodes that do not work. In comparison, surface emitters can be tested during the production cycle, which can result in problems being recognized quicker. However, because of the high mirror reflectivities, VCSELs have a lower output power when compared to edge-emitting laser diodes.

6.1.4.6 Vertical External Cavity Surface-Emitting Laser   In concluding our brief discussion on different types of laser diodes, we will turn our attention to vertical external cavity surface-emitting lasers (VECSELs). A VECSEL is similar to the previously described VCSEL, with the latter having its mirrors commonly grown epitaxially as part of the diode structure or bonded directly onto the semiconductor material containing the active region. In comparison, VECSELs are fabricated so that one of its two mirrors is external to the diode. This fabrication results in the cavity having a free-space region.

VECSELs can be either optically or electrically pumped. Optically pumped VECSELs are commonly used in such high-power applications as industrial cutting, whereas electrically pumped VECSELs are commonly used in projection displays.

6.2 Comparing Laser Diodes and LEDs

Now that we have a general appreciation for several types of laser diodes and their basic operation, we have a conceptual foundation we can use. Thus, in this section, we will compare and contrast many of the key operating characteristics of each device.

6.2.1 Comparing Operational Characteristics

Although the output of laser diodes in the form of coherent light represents a major difference between lasers and LEDs, there are other differences that govern the applications for which each device is

Table 6.1 Comparing Laser Diodes and LEDs

CHARACTERISTIC	LASER DIODE	LED
Light output	Coherent	Noncoherent
Output power	Proportional to current above the threshold	Linearly proportional to drive current
Current	5–40 mA threshold current	50–100 mA drive current
Coupled power	High	Moderate
On/Off speed	Higher	Slower
Bandwidth		
Use with fiber	Single mode, multimode	Multimode
Spectral width	0.00001–10 nm	40–190 nm
Wavelengths	0.78–1.654 nm	0.66–1.65 nm
Bandwidth	Higher	Lower
Longevity	Long	Longer
Cost	Higher	Lower

commonly used. Table 6.1 provides a general comparison of 12 characteristics of laser diodes and LEDs. Although the entries for some of the characteristics listed in Table 6.1 can be considered to be subjective, this results from the fact that we are comparing a range of LEDs against a range of laser diodes such that a subjective response to general characteristics of each device is warranted.

6.2.2 Performance Characteristics

When we describe and discuss the use of laser diodes in different types of applications, their usefulness is governed by one or more of the performance characteristics of the laser. Such characteristics, which are extremely important, especially, for high-speed data communications, include their turn on/turn off speed, peak wavelength, power, spectral width, emission pattern, and linearity. When all these characteristics, which we will shortly describe, are considered, we will note that they provide the capability to be employed with fiber optics to support Gigabit and 10 Gigabit Ethernet network construction and operation.

6.2.2.1 Speed   Lasers provide a high-speed turn on/turn off capability as their rise and fall times are relatively short in comparison to LEDs. This enables laser-based communications to occur at data rates beyond 622 Mbps, which is the highest operational rate currently supported by LEDs. Thus, the pulse rise and fall times of lasers are shorter than

LEDs, enabling lasers to support very-high-speed communications that are currently beyond the capability of LEDs.

6.2.2.2 Peak Wavelength   Peak wavelength, as the term implies, represents the wavelength at which a source emits the most power. Ideally, the peak wavelength should be matched to the wavelengths that can be transmitted through fiber cable with the least attenuation.

For laser diodes, the most common peak wavelengths are 1310, 1550, and 1625 nm. This explains why laser diodes commonly operate at 1310 or 1550 nm over single-mode fiber or Gigabit Ethernet networks commonly referred to by their IEEE nomenclature as 1000BASE-LX and 1000BASE-LX10. In general, laser diodes have a slightly narrower range of wavelengths than LEDs.

6.2.2.3 Power Coupling   Power coupling represents the amount of power from a laser diode or an LED that is transmitted into a fiber-optic cable. Obviously, the higher the percentage of power that can be coupled into the fiber, the better the transmission possibility. What is desired is the output power of the source that flows through the fiber should be of sufficient strength so that it can provide a sufficient amount of power to a detector located at the opposite end of the cable. Lasers in general have a higher power coupling than LEDs.

6.2.2.4 Spectral Width   If a laser performs perfectly, all of its light emission would occur at the peak wavelength. Unfortunately, lasers similar to humans have imperfections, and the light emitted by a laser occurs in a range of wavelengths that is centered around the peak wavelength. The range of wavelengths is referred to as the *spectral width,* and the narrower it is, the more focused the emitted light. Lasers in general have a much narrower spectral width than LEDs.

6.2.2.5 Emission Pattern   The amount of light that can be coupled into an optical fiber is governed by the pattern of emitted light a laser diode generates. For best results, the size of the emitted region of the laser diode should very closely match with the diameter of the core of a fiber-optic cable.

When comparing laser diodes to LEDs in terms of their emission patterns, there is no contest. The emission spectrum of an LED is much broader than that of a laser diode. This results from the fact that laser light output is more focused than an LED. Thus, lasers can be coupled to single or multimode fiber, whereas LEDs can be coupled only to a multimode fiber.

6.2.2.6 Linearity *Linearity* represents an important characteristic of both LEDs and laser diodes in many applications. This term defines the relationship between the optical output of an LED or laser diode and its electrical current input.

In a perfect world, the linearity of a laser diode would be a straight line that plots optical power in milliwatts against the forward current expressed in milliamperes. However, because lasers are temperature-sensitive devices and, similar to LEDs, have an initial poor slope efficiency, the light output of a laser diode is both partially nonlinear and varies the basis of temperature.

Similar to LEDs, a laser diode has a threshold current that represents the minimum current necessary to activate the device. Thus, until some forward current level is reached, the optical output power of a laser diode is zero. As the operating temperature of a laser diode changes, this results in a change to the threshold current, with a higher temperature resulting in a higher threshold current. Another change associated with the operating temperature of a laser diode is its slope efficiency, with most lasers exhibiting a drop in slope efficiency as the operating temperature increases.

To achieve a maximum level of performance, lasers are often used with a photodiode to provide a feedback control mechanism. This is accomplished by monitoring the light output on the rear facet of the laser. Because the current from the photodiode varies in tandem with the light emitted, its feedback can be used to adjust the laser drive current.

Figure 6.7 illustrates the effect of temperature on the optical output of a laser diode. In this example, $T_1 < T_2$ and the curves are generic and do not represent a specific laser diode. Instead, the curves are used to illustrate the effect of a change in operating temperature on the optical output of a laser diode. As indicated in Figure 6.7, as the temperature of the laser increases, the curve is shifted to the right.

When comparing laser diodes to LEDs, both devices exhibit a similar linearity above a threshold current. However, if you were to compare power output versus forward current curves from different

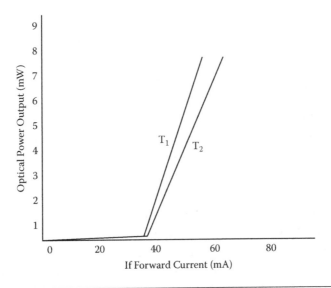

Figure 6.7   Effect of operating temperature on laser diode optical power output.

manufacturers, you would more than likely note that the laser diode curves have a more pronounced slope.

6.2.2.7 Luminous Efficacy   Commercially available laser diodes can be obtained with an overall conversion efficacy between 45 and 50 percent, with laboratories and research centers examining methods to enhance the overall efficiency to 75 percent or more. To compare LEDs to laser diodes, we need to consider a variety of color LEDs. Some LEDs, such as bright red, achieve a luminous efficiency (lumens output per watt input) of 20–40 percent, with other colors having a luminous efficacy ranging from a low value of 5 percent for certain yellows to 12 percent for certain blues and near 20 percent for different oranges. Although research into improving LED efficacy is progressing at a rapid pace, currently laser diodes are more efficient than a typical LED. However, within a few years, improvements in LEDs used for lighting should enhance their efficiency to a considerable degree.

6.2.2.8 Drivers   Both laser diodes and LEDs can be obtained as stand-alone devices or integrated in a package. Concerning the latter, a typical package includes driver circuitry that protects the device from too much current. When obtained as a package, you typically can connect the package to a dc power supply or battery to obtain a fixed output, or use a capacitor and a resistor to obtain a variable output.

6.2.3 Safety

When comparing laser diodes to LEDs in terms of safety, we can make an intelligent guess that laser diodes are more dangerous. After all, lasers are used in surgery, evaluated for missile defense, and they function as a tool for precise cutting of a variety of fabrics to different types of metal. In comparison, LEDs can be considered as a more passive cousin, primarily used for illumination.

Instead of making an intelligent guess, we can examine the vision safety of each device scientifically. We can define the intensity of a light source as the optical power produced divided by the area of emission, with the result divided by the angle of emission. Both LEDs and laser diodes that are obtained with the same power output and have the same angle of emission have considerable differences in their emitter area. An LED will emit light from the surface of its chip over a full hemisphere. However, a lens is added to radiate light over a limited angle of viewing. This lens does not change the LED's emitting area, which is approximately 0.3 mm × 0.3 mm or 0.09 mm². In comparison, a typical laser diode has an emitting area approximately equal to 1 μm × 3 μm or 3 μm². Thus, without considering the use of an LED lens when you compare the emitter area of an LED to a laser diode, the LED's emitter area is larger by a factor of 30,000! This means that when you compare an LED and a laser diode having the same light intensity and consider the emitter area, the laser intensity can be approximately 30,000 times that of an LED, which is the reason you are told not to look directly toward the light emitted by a laser-based pointer.

Because the optical power of a laser is in effect focused whereas that of an LED is spread out, the laser can actually burn a tiny area of your retina. In fact, laser diodes are now commonly used in many types of eye surgeries, ranging from vision correction to repairing tears in the eye.

6.2.4 Applications

As briefly discussed in the beginning of this chapter, the LED and laser diode applications can be complementary or dissimilar to one

Table 6.2 Comparing LED and Laser Diode Applications

APPLICATION AREA/APPLICATION	LED	LASER DIODE
Commercial		
Paper check/motion detection	X	
Pointer	X	X
Data communications		
Low speed	X	X
High speed	X	X
Dental drill		
Fiber-optic tooth illuminator	X	X
Illumination		
Low voltage lighting	X	
General lighting	X	
Medical		
Cosmetic surgery		X
General surgery		X
Military		
Lighting	X	
Range finder		X

another. In this section, we will use the previously described characteristics of both devices to discuss some of the applications they are well suited for.

Table 6.2 lists six application areas where we will review the use of laser diodes and LEDs. For each area, we will discuss one or more applications and how one or both devices are suitable or not suitable to be used for a specific application.

6.2.4.1 Commercial Applications In the commercial application area, this author lists paper check/disbursement and pointer as two representative applications that can be used to define differences between the use of LEDs and laser diodes. The paper check is an LED activity in which a break in the light is used to define an "out of paper" printer condition. Although a laser diode could be used for this application, it's more expensive and has a potential to adversely affect a person's vision if a person views the light directly. Thus, the LED dominates this application area.

Moving our attention to the motion detection area, LEDs can be used to provide a light barrier that when broken results in some activity taking place, such as a paper towel being dispensed or an alarm being activated. Although a laser diode could be used for motion detection, in most commercial applications, it is the LED that is used to provide the light source that when broken triggers an action. This is again due to safety issues. Perhaps, a key exception is the use of lasers to provide an in-building motion detection system that monitors a predefined area and if the light is broken, triggers an alarm and perhaps the movement of the item or items protected into a more secure area or the locking of the room where the items are located. Some examples of the use of laser diodes for motion detection include museums and art galleries. Because the laser can be used to provide a narrow beam of light, several laser diodes can provide a "cross-hair" pattern that makes it impossible to cross a barrier without breaking a light beam and generating an alarm. Typically, movies featuring an art heist show good examples of the use of laser diodes as security measures within a museum. Although most art heist movies show the burglar in one manner or another overcoming the laser diode system and leaving a museum or art gallery with their illicitly obtained artwork, in reality it is much more difficult to do so. In fact, museums and art galleries using a laser diode protection scheme are rarely compromised.

6.2.4.2 Data Communications Both LEDs and laser diodes play an important role in the field of data communications. As previously noted in this chapter (Section 6.2.2.1), LEDs can provide a maximum data transmission rate of 622 Mbps due to their pulse rise and fall times, which currently preclude a faster modulation rate with existing state-of-the-art technology. Thus, ultraviolet LEDs are used to extend Ethernet at 10 Mbps and Fast Ethernet at 100 Mbps, and other types of LANs as well as OC-3 (optical carrier) and OC-112 intracity communications at data rates of 155.52 and 622.08 Mbps. In comparison, the faster pulse rise and pulse fall times of laser diodes make them suitable for use in Gigabit Ethernet and 10 Gigabit Ethernet LANs. In addition, because laser diodes can support the use of single-mode fiber, they not only can transmit further than LEDs but in addition enable repeaters to be spaced further apart than when LEDs are used and fiber is routed between cities. This in turn lowers the cost of communications as the number of repeaters is reduced.

6.2.4.3 Dental Applications Although this author doubts if readers like to visit their dentist, both LEDs and lasers can be found in some offices. Because the laser can have enough focused coherent light to burn bone, it is being used as a drill substitute in some dental offices. In comparison, the LED is used to generate light into a fiber-optic bundle that surrounds a conventional dental drill, illuminating an area around the drill bit that helps a dentist view their drilling activity. Thus, the first application takes advantage of the power of the laser, whereas the second takes advantage of the illumination capability of the LED.

6.2.4.4 Illumination Application Although there has been mention of the laser light bulb in several publications, it represents a situation in which technology has not matured into reality. Today, we are witnessing the replacement of incandescent lighting by compact fluorescent lighting (CFL) and the gradual replacement of both incandescents and CFLs by LED-based lights.

Low-voltage lights or solar power lights used in gardens that require no electrical cord are usually based on the use of LEDs. Similarly, conventional lighting for hard-to-reach locations that require long-lasting bulbs to minimize maintenance expenses are now almost all LED-based lightbulbs. As economies of scale bring down the cost of LED-based lightbulbs, we can reasonably expect their extremely long life and low energy consumption to make them very attractive candidates for the replacement of existing incandescent and CFL lighting.

6.2.4.5 Medical Application Once again, the ability of lasers to burn through everything from fabric and metal to skin in a controlled manner makes them very suitable for a variety of medical procedures ranging from general cosmetic surgery, such as hair or wrinkle removal, to laser eye surgery. In comparison, the inability of LEDs to provide sufficient power and focused light precludes their use in surgical applications.

6.2.4.6 Military Applications There are certain properties of both LEDs and laser diodes that make them very suitable for different types of military operations. Concerning LEDs, as a solid-state device they can be exposed to a considerable level of shock and still operate. This can be very important when considering interior lighting within vehicles such as tanks and armored personnel carriers as well as certain types of flashlights used for reading military maps. Until recently, lightbulbs manufactured to military specifications were primarily used in military vehicles. In addition to being rather costly, their life expectancy was relatively short. Through the use of LED-based lightbulbs, interior lighting of many military vehicles are able to withstand significant shock and vibration, last longer, consume less power, and they actually cost less than incandescent lightbulbs manufactured to military specifications.

If we turn our attention to flashlights, we can note that the military, like civilians, are now purchasing LED-based devices. Not only are such devices brighter, but most importantly their lower level of power consumption provides less drain on the batteries, enabling a doubling or tripling of time between battery changes in comparison to the need to change batteries frequently in a conventional flashlight. Thus, LEDs are emerging as components of various types of military illumination requirements.

Perhaps the first use of lasers by the military was as a range finder for the main weapon mounted on a tank. By measuring the time until a reflection occurs from the narrow beam of laser light used to "paint" a target, it becomes possible to determine the range to the target. This information is then used by the on-board computer to adjust the angle that the main gun uses to fire its projectile.

Since its use as a tank range finder, the laser has been incorporated into a variety of weapon systems. During both the first and second Gulf Wars, the U.S. Air Force heavily relied on guided munitions dropped by one aircraft that followed a laser "track" to a target illuminated by a second aircraft. Due to the inability of lasers to be jammed, they are ideal for use with precision guided weapons as long as cloud cover does not obscure the illumination. In addition to serving as a guide for the delivery of precision-guided munitions, other types of lasers are being used in prototype systems designed to intercept both ballistic and short-range missiles. Once again, it is the ability of lasers to generate a powerful focused beam of light that makes them suitable for such applications and distinguishes their use from that of LEDs, which are more suitable for illumination applications.

7

THE EVOLVING LED

No book focused on an evolving technology would be complete without considering where we might be going with respect to future technology. In this chapter, this author looks into his crystal ball and attempts to predict a few fresh developments that can be expected to occur in the development cycle of light-emitting diode (LED) technology. Some of the topics we will discuss in this chapter were previously mentioned in this book, whereas other topics may be pure speculation based on educated guessing by this author. Regardless of the topic, the intention of this chapter is to make readers aware of the growth potential for the use of LEDs. That growth potential is envisioned by this author in three key areas: lighting, communications, and display.

7.1 Lighting

In this book we carefully examined the present state of LED-based lighting. In doing so, we noted that the LED lightbulb has the potential to replace compact fluorescent light (CFL) in approximately the same manner as the CFL lightbulb is replacing the incandescent lightbulb. However, when we discussed the evolution of the LED lightbulb, we primarily focused our attention on its extremely low power consumption, and its relatively low output of light in comparison with other lighting sources. In this section, we will focus our attention on the latter and note how LED-based bulb manufacturers are attempting to provide more light output per bulb.

There are basically two methods being used by LED bulb manufacturers to increase lighting. Those methods are increasing the LED density and the light output per LED.

7.1.1 Increasing LED Density

Although it may appear obvious that engineering a lightbulb with additional LEDs will produce more light, it is anything but a simple task to do so. Thus, as the density of LEDs increases, the cost of the bulb, as expected, also increases. Recently, several manufacturers have been able to make a significant leap forward, introducing products with 80 to 100 or more LEDs in a lightbulb package in which the previous high was between 30 to 36 LEDs. Thus, the number of LEDs per bulb has recently tripled. Unfortunately, a survey of Internet sites selling LED-based lighting indicates that the cost of the more densely populated lighting has increased by a factor between 8 and 12. Thus, although LED density has increased, the retail price of the bulbs has risen even more.

The relatively low power consumption of LEDs will make them more attractive on a lifetime basis as the cost of electricity continues to rise. Thus, the additional demand for LED-based lighting can be expected to increase the manufacturing of such lighting. This in turn can be expected to increase economies of scale, resulting in an eventual decrease in the unit cost of LED-based lightbulbs. Thus, within a few years, this author expects high-density LED-based lightbulbs to become very price competitive with CFL lightbulbs. In addition, LED-based lightbulbs have a significant advantage over CFL with respect to their environmental impact. As previously mentioned in this book, CFL bulbs contain mercury, which will eventually result in a significant landfill problem as more and more CFL bulbs reach the end of their life and have to be disposed of. In comparison, LED-based bulbs do not have this problem.

7.1.2 Light Output per LED

Currently, the primary method used by LED bulb manufacturers to increase their lumens or light output is to pack more LEDs into a bulb. Although significant progress has been made by developing ultrabright LEDs and incorporating them into lightbulbs via densely packed technology, the total lumens per bulb currently is below 600. In comparison, a 26 W pin-based CFL bulb produces 1800 lm, whereas a common 15 W Edison-type base CFL bulb produces 750 lm of light. Thus, in the general illumination area, LED-based lightbulbs appear to have a considerable need for additional light output to be competitive with incandescent and CFL lighting. In actuality, due to the density of LEDs in many lightbulbs, only a small advance in lumens per LED is necessary to meet or exceed the lumen output of other types of lightbulbs.

It is this author's opinion that advances in ultralight LED technology will result in the availability of LED-based lightbulbs that will match or even exceed the light produced by other types of lightbulbs. This opinion is based on research being performed in the United States, Europe, and Japan to increase the lumen output of LEDs. Over the past few years, significant progress has been made, resulting in the development of ultrabright LEDs that have reached the market. If continuing progress occurs at only half the rate of success of prior years, this author expects LED-based lightbulbs with over 750 lm to reach the market by 2010, with 1000 lm bulbs to follow within a year or two. In the interim, due to the highly directional nature of light generated by LEDs, they are becoming highly competitive for under-cabinet and other niche lighting applications in which the directional nature of LED lighting is highly advantageous over incandescent and CFL, which basically generate light in all directions. In effect, the typical under-cabinet fixture designed for use with a halogen or fluorescent light is between 30 and 50 percent efficient, which means that approximately half of the light never leaves the fixture. Similarly, the use of CFL or incandescent lightbulbs in recessed ceiling fixtures may result in 50 percent or more of available light being lost. Thus, both under-counter and recessed lighting as well as focused track lighting represent areas where the lower light output but highly directional nature of LED lighting can result in viable products. In fact, a quick search by this author using such terms as "LED track lighting," "LED under-cabinet lighting," and "LED recessed lighting" turned up a number of companies selling such products. Although these are currently niche products due to the relatively high cost of LED lightbulbs, within a few years, the previously mentioned advances in LED technology will have a significant effect on the mass consumer market for such lightbulbs. To illustrate why this will occur, consider the plot of electrical consumption versus brightness for incandescent, halogen,

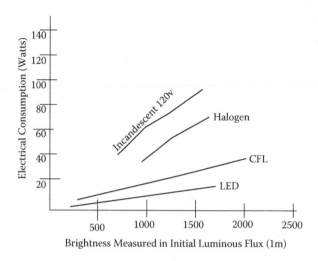

Figure 7.1 Electrical use by bulb type.

compact fluorescent, and LED-based lighting shown in Figure 7.1 that should be available within a few years. Although LED lighting may not provide a very high level of brightness as compared with halogen and certain CFL, for a majority of applications, it can be expected by 2010 to represent a preferred method of lighting due to its extremely low electrical consumption and significant range of brightness levels.

The major advantages of LED-based lighting are similar to those of CFL over incandescent: they do not generate heat similar to incandescent lightbulbs; they last longer than incandescent and CFL-based bulbs; they save energy consumption, and perhaps, most important when discussing energy efficiency, the LED-based bulb unlike a CFL bulb does not contain any toxic chemicals. Because CFL bulbs contain mercury, their breakage inside the home or office can require a considerable cleanup effort as well as the evacuation of pets and children from the area where the breakage occurred. In addition, as CFL bulbs burn out and are trashed, a future problem will arise because they will leak mercury into landfills. Thus, a recycling program will eventually be needed for CFL, which will drive up its cost. In comparison, LED-based lightbulbs are mercury free, and their long-term effect in landfills should be minimal in comparison with CFL.

7.2 Communications

The field of data communications represents a second area of technology where this author believes the use of LEDs can considerably expand. Although LEDs are presently limited to a modulation rate of 622 Mbps, which makes its use unsuitable for Gigabit Ethernet and 10 Gigabit Ethernet networking, it's important to remember that the data rate is a function of both the modulation rate and the number of bits packed into each signal, technically referred to as a *baud*. Although the current use of LEDs for communications results in digital data converted into light pulses on a one-to-one basis where the bit rate is then equal to the baud rate, the data rate can be improved by packing more bits into each modulated pulse. For example, using a pulse width modulation (PWM) scheme, the pulse can be enabled for different durations as illustrated in Figure 7.2.

Thus, the duration of the pulse can be used to denote a sequence of bits. For example, consider the entries in Table 7.1, which list a possible dibit pulse modulation scheme. Here, two bits are used to generate a specific pulse width.

Table 7.1 Potential Dibit PWM Scheme

BITS	PULSE WIDTH
00	0
01	1
10	2
11	3

In examining the entries in Table 7.1, note that the absence of a pulse width would reflect the bit pair 00. In comparison, pulse widths of 1, 2, and 3 would reflect bit pairs 01, 10, and 11, respectively. As two bits are packed per baud, if a signaling rate of 622 Mbps is maintained, the data rate becomes 622 Mbps × 2 or 1.244 Mbps, which is more than sufficient for Gigabit Ethernet.

The ability of LEDs to provide a higher data transfer rate will depend on photodetectors that can measure pulse widths. Within a few years, this author believes that advances in technology will make this a reality, allowing LEDs to support higher-speed communications beyond 622 Mbps.

Figure 7.2 Pulse width modulation.

7.3 Organic LEDs

Organic LEDs (OLEDs) perhaps offer the brightest capability, no pun intended, for the future use of LEDs. Although OLEDs are primarily considered as an evolving mechanism to increase the readability and durability of displays as well as decrease their power consumption, a second application area is emerging that offers the potential to considerably enhance the growth in the use of OLED technology. That application area is lighting.

In this concluding section of this chapter, we will briefly discuss the use of OLED technology in displays and as a lighting mechanism. As we discuss the use of OLED, we will note some of the major advantages of the technology, which will be a driving force for its adoption into consumer products.

7.3.1 Display Utilization

Earlier in this book, we noted that Sony was actively marketing an OLED-based television, whereas several other manufacturers had developed prototype systems. Although the technology behind OLED displays is chemical, the applications that can use the technology range in scope from television screens and computer displays to billboards, cell phones, stereo displays, and even navigation systems. In effect, any display represents a potential OLED application. However, costs, life of certain colors, and other factors currently act as constraints in rapidly migrating OLED technology to applications.

7.3.2 Advantages

From a manufacturing perspective, only a limited number of steps are required to fabricate an OLED display. In fact, an entire display can be built on a sheet of glass or even plastic, which considerably reduces the manufacturing cost. From the consumer perspective, OLED-based displays allow all colors of the visible spectrum to be shown, and have a high brightness that is particularly useful when watching television or working with a notebook computer in a sunny area. In addition, an OLED display has no viewing angle dependence unlike plasma and LCD displays, and has a high pixel response rate that removes blurring from sports and gaming action that commonly occurs on plasma- and LCD-based televisions, and computer displays.

7.3.3 Current Deficiencies

Although there are considerable advantages for both the manufacturer and consumer, currently OLED technology has a number of hurdles that need to be overcome. The first key hurdle is display life with 10,000 hr, now suitable for cell phones and notebooks but relatively low for a large-screen television that represents a considerable one-time purchase cost. A second hurdle involves the fact that red, green, and blue emitters degrade at different rates, with blue sometimes having a half-life of approximately 4,000 to 6,000 hr. In recognition of these two hurdles, vendors began looking for methods to extend the life of the display as well as each of the three primary color emitters. Recently, the lifetime for blue was extended to approximately 30,000 hr, with much longer periods for red and green that approach or exceed a hundred thousand hours for each color. As this technology is incorporated into consumer products, the superior color, viewing angle, and higher pixel-painting speed should enable OLED displays to move from a niche market into a viable mainstream commercial market.

One of the key attributes of OLED technology is its ability to use plastic, making it possible to have a roll-up, large-scale television one day. However, prior to roll-up, plastic-based televisions becoming a reality, the problem of exposure to water or oxygen needs to be successfully addressed.

OLED displays incorporate chemicals and metals that can be ruined by exposure to water or oxygen. Because glass is virtually impervious to water or oxygen, this is not a problem for rigid OLED displays. Unfortunately, plastic is porous and needs to be coated with a barrier to prevent water and oxygen penetration that can ruin an OLED-based display. Although current technology has not produced a barrier that is flexible and transparent enough to enable a roll-up display, this author believes it's just a matter of time until the correct set of chemical coatings is found. Once this occurs, roll-up OLED plastic televisions will move from science fiction to reality.

7.3.4 Lighting

Although not thought of as a potential replacement for CFL and incandescent lighting, OLED technology provides the ability to generate light much more efficiently than current incandescent or fluorescent lighting. By combining two layers of phosphorescent diodes to release green and red wavelength lights with an additional layer of a fluorescent diode to provide blue wavelength light, the three layers produce very efficient white light. As currently in the research stage, the work of Stephen Forrest of the University of Michigan and Mark Thompson of the University of Southern California is being expanded, and within a few years, it might be possible to have OLED-based lighting built into walls, furniture, and even sliding doors and windows. The key to this possibility is the fact that the organic layers are only 10 nm thick and transparent when turned off. Although the layers of diodes can be fabricated using glass or plastic, it appears that plastic would be better due to its flexibility; however, the plastic will require a moisture barrier that will add to its cost.

Because the cost associated with manufacturing a lightbulb is minimal, this author expects OLED lighting when commercialized to initially target niche markets. Eventually, with the demise of incandescent lighting due to legislative action, it may be possible for OLED lighting to expand its use to other residential and commercial lighting applications. Thus, in closing, we can say that the future of LEDs is bright and can be expected to be even brighter.

Index